AI FOR LUNCH

A PLAIN-ENGLISH GUIDE TO HOW AI ACTUALLY WORKS, AND WHEN TO TRUST IT

AI AT THE TABLE
BOOK 1

THEO KERNEL

Published by Vector Verse Pty Ltd, Australia.

First edition, 2026.

This book explains real technology through an invented setting. The kitchen, its cooks, and every establishment in these pages are inventions; no real person, company, or product is named or depicted. How the book was made, and why nothing real is named, is described in "A Note on Names, and on How This Book Was Made."

This book provides general information about artificial intelligence. It is not professional advice, and nothing in it is legal, financial, or medical advice.

Paperback ISBN: 978-1-7648284-0-6

Vector Verse Pty Ltd · www.vectorverse.com.au

A NOTE ON NAMES, AND ON HOW THIS BOOK WAS MADE

No company, product, model, lab, or famous name appears anywhere in these pages, and no dates. That is deliberate. A book that names this year's tools is out of date by next year's, and a book that leans on famous voices is borrowing authority rather than earning it. Nothing here is named, so nothing here can go stale and nothing here is owed. What is left is the only standard worth accepting: every claim is about how the kitchen works, put plainly enough that you can test it at your own counter.

Same kitchen, same rule, all the way down.

YOU ARE EATING FOOD YOU NEVER SAW COOKED

Here is something strange you have started doing, probably in the last year or two, almost certainly without noticing how strange it is. You ask a machine a question, in plain words, the way you would ask a friend who happens to know things. It answers in plain words, instantly, fluently, with complete confidence. And you have no idea how it did that, whether the answer is true, or what, precisely, you are talking to. You take the plate. You eat. You come back for more.

I am not going to scold you for this, because I do it too, and so does everyone reading over your shoulder. I will take you up to the kitchen, because the gap between how much we now lean on these systems and how little we understand them is the widest I know of, and an uncomfortable place to be standing.

Here is how you meet one of these things. You judge it the way you judge any meal: good, wrong, better than I expected. What you never see, and what almost no one will explain to you without either selling you something or trying to frighten you, is the kitchen on the other side of the wall.

This book is a seat with a view through to that kitchen, and it runs on a single bet: in a world filling up with these machines, the people who do best will not be the ones who can build them. Very few people can, and you are probably not about to join them. The people who do best will be the ones who can read them, who can look at a confident answer and know when to trust it and when to send it back. The ones who cannot are the ones who quietly pay for it: the invented citation that goes out in a brief under their name, the answer trusted precisely because it looked the part. Not one of those mistakes announces itself.

Follow one through. Someone needs the standard lead time on a piece of work, asks, and back comes a number: clean, plausible, specific to the day. It goes straight into a reply to a customer, because it looks like every lead time anyone has ever quoted, and lead times are the sort of thing somebody else has already confirmed. The customer says yes in writing. It is still sitting in that thread the following week, when the person who actually schedules the work reads it.

The skill is worth having before you need it. That skill needs no mathematics. It is the one this book hands you.

By the last page you will be able to say out loud, without bluffing: what one of these systems actually is, which is neither a brain nor a robot but something stranger and more ordinary than both; how it learned to do what it does; why it will tell you something false in the exact calm voice it uses for the truth; how to catch it; and how to get a sharply better result out of it than you are getting now. That last one alone is worth the lunch.

Why a kitchen, and not a brain? Because "brain" makes you picture understanding that is not there, and "robot" makes you picture a will that is not there, and each will steer you wrong on the day it matters. A kitchen makes you picture what is true: raw material in, finished product out, running on ingredients and equipment and skill, with a rhythm and a short list of predictable ways to fail. You already know in your body that bad ingredients sink a brilliant cook, that a kitchen buried in orders starts cutting corners, that a place that has only ever cooked one cuisine will fumble the first unfamiliar order. Every one of those instincts is about to go to work.

One promise about the method, and the whole book depends on my keeping it. An analogy simplifies; that is its job. This one will not lie to you. Every time the kitchen starts to bend the truth about the real machine, I will stop, put a finger on the exact spot, and tell you where the picture tears. An explanation that hides its own seams is not teaching you — it is flattering you, and you can get flattery for free anywhere. Two ideas will carry more weight here than any others: always taste before you serve, and treat the kitchen as a prep cook rather than a replacement. You will meet them again at a different altitude in each book of this series.

So take the best seat in the house for this, the one at the counter by the window to the back: close enough to feel the heat, not close enough to get burned. And here is the strangest part of it, the part you have been doing this whole time without a second thought. You have been trusting the cook, and you have never once seen the kitchen.

ORDERING FROM A KITCHEN YOU CAN'T SEE

Something happens when you order at a busy counter place, the kind with a narrow window between the dining room and the back. You step up. You state what you want, or write it on a slip and push it across. The slip disappears. For a stretch, nothing. You hear the scrape of a pan, a low voice calling something you cannot make out. Then a plate appears on the ledge with your number on it, and a finished meal is yours. You never saw the burner. You never saw whose hands made it. You saw a request go in and a result come out, and everything between happened behind a wall.

Hold that picture. When you use one of these AI systems, that is the whole transaction from your side. Something goes in. Something comes out. The making is behind the wall, and the making is the part you need to understand.

The thing behind the wall has a name. It is called a model. A model is the trained thing itself, built to take an input and turn out an output of a particular kind. Read that slowly, because every word earns its place. Trained, which the next several chapters unpack in full, means it learned from examples rather than being handed a rulebook. An input and an output of a particular kind means each kitchen is built for a certain sort of order. A kitchen that turns words into words sits behind most chat tools. Others turn a description into a picture, or speech into text, or your recent behavior into a recommendation. Different produce, meaning the pile of examples each one learned from, and different equipment. The same fundamental shape: an order in, a made thing out.

So far this could describe a vending machine, and that is the wrong picture. Almost every serious misunderstanding of these systems begins here. A vending machine has the candy bars already inside it. You press B4 and a specific, pre-made object drops. The kitchen does not work that way. There is no shelf of pre-written answers waiting for you to pick one. It cooks your plate when you order it, from scratch, every time. Ask the same question twice and you can get two different answers, the way the same cook turns out two slightly different omelets on two mornings. There is no drawer of finished omelets anywhere in the building. There is a cook who knows how to make one.

That single fact resolves most of the confusion. A model does not look things up. You ask a question, a fluent answer comes back, and it feels as though something reached into a vast filing cabinet, found the right page, and read it to you. That is not what happened. The gap between what happened and what it feels like is the source of more costly mistakes than anything else in this book. The kitchen did not retrieve a finished dish. It cooked one to order, drawing on everything it learned in training about what a good answer to a request like yours tends to look like. Think of a cook who has tasted so many thousands of versions of a dish that she can produce a convincing one on demand. Not a clerk fetching the right file. Mark this one, the first tear in the picture: the cook has a palate. Nothing in the machine does. What stands in for it is a set of numbers settled during training, and the resemblance ends there.

Insist on the distinction, because a cook producing a convincing plate and a clerk fetching a correct file fail in different ways, and if you expect the clerk when you have the cook, the failures will blindside you. The clerk's errors are errors of retrieval. The file is missing, or it is the wrong file. The cook's errors are errors of plausibility. The dish looks right, is plated with care, and is nonetheless not the dish you ordered, because the cook was producing something convincing rather than consulting something true. We give that failure its full and dangerous due later, when the kitchen burns. For now, fix this in place and do not let it move: the kitchen makes things, it does not retrieve them.

I believed this in the abstract before I really believed it. What moved it was a demonstration I watched: the same question put to the same system twice, once on its own and once after it was handed a set of source documents to work from. Two answers came back. The first was fluent and confident and assembled out of the kitchen's general sense of how such an answer goes. The second was narrower, plainer, and tied to the documents in front of it, still cooked rather than quoted. Nothing about the kitchen had changed in between. The only difference was whether it had anything real to cook from. Seeing them side by side is what made the distinction stop being a sentence I agreed with and start being a thing I could recognize. How that second arrangement works is a later course. What it showed me on the spot was what the first one had been doing all along.

One more distinction will save you from a confusion that muddies nearly every news story about this technology. It is the difference between the kitchen and the restaurant. When you use an AI product, you are not touching the model alone. You are touching a whole restaurant built around it. A host who greets you. A menu that suggests what to order. A dining room with a certain decor. A waiter who remembers you came in last week. A runner who can dash next door for something the kitchen does not stock. All of that is real, and all of it shapes your experience, and none of it is the kitchen. The kitchen is the model, the part that does the actual cooking, that takes the raw order and produces the made thing. Everything else is the restaurant around it.

Keep that line sharp and much of the noise clarifies. You read that a system can now remember your past conversations. That is almost never a change to the cook. It is the front of house keeping a notebook on regulars and reminding the cook what you usually order. You read that a system can now search the web or run a calculation. That is the restaurant hiring a runner to fetch from outside, not the cook gaining a new skill, though deciding when to send him becomes part of the cook's job, and no more reliable than the rest of it. We have whole chapters coming on the notebook and the runner. The only reason to separate them now is that the most confused arguments about these systems come from mistaking something the restaurant did for something the kitchen is. The kitchen is one specific thing. The restaurant is everything wrapped around it, and the two change for different reasons.

There are two errors waiting here, and both of them are generous to the kitchen, which is what makes the subject treacherous: escaping the first is no protection against the second. The first error is to treat the kitchen as a mind. The plates are so polished, the language so fluent and human, that it becomes almost irresistible to assume someone back there means it, knows what is true, wants to help you, or now and then wants to mislead you. No one does. There is a cook producing plausible plates, with no beliefs about your question and no intentions toward you, the way a line cook has no feelings about the four hundredth order of the night. Read intention into the kitchen and you will trust it in ways it has not earned, and argue with it as though it could be persuaded. Both are wasted motion.

The second error is the more dangerous of the two. It is the error of treating the kitchen as an oracle, of assuming that because a plate looks expert, it is correct. The kitchen's product is plausibility. It is extraordinarily good at producing a dish that looks like the right answer. But looking like the right answer and being the right answer are different things, and the distance between them is where careful, intelligent people get taken in, because the kitchen gives no signal that you have crossed from one to the other. A beautiful plate can be the wrong dish. It will not look any less beautiful for being wrong.

Hold both errors in view and you arrive at the right relationship with the kitchen. It is not a person; there is no mind back there to look for. And it is not an oracle, so a pretty plate is not thereby a true one. What it is, is a remarkably capable cook that makes convincing food to order. Your job, as the one eating, is to stay the kind of diner who can tell the difference.

Carry one picture out of the dining room. A confident answer tells you nothing about whether it is true, because fluency is the kitchen's default tone, worn over fact and fiction alike. When an answer appears, ask the one question that holds all of this together: did the kitchen cook this, or was it fetched? It cooked it. It has nothing to fetch from; fetching is the runner's job, out in the restaurant, and whatever the runner brings back gets cooked all the same. So treat even a polished plate as convincing rather than confirmed, until you have checked the part that matters.

TASTE IS LEARNED, NOT PROGRAMMED

COOKING BY THE RULEBOOK

A kitchen from a few decades back runs from a binder. A thick one. Its cover gone soft and gray from handling, it sits open on a stand at the center of the line, and every dish the kitchen can make is in there, along with a rule for every situation a cook might face. If the bread is this shade of brown, it is done. If a customer asks for the soup, ladle exactly this much. Asked whether a dish is spicy, the cook says these precise words and no others. The cooks are fast. They are tireless. And as long as a ticket matches something in the binder, the plates come out clean and identical, every single time, with a consistency no human cook could ever match.

Then a ticket arrives for something the binder does not cover. A small substitution. An odd request, phrased in a way no one anticipated. The line stops. The cooks flip back and forth through the binder, and there is simply no rule for this, and they cannot improvise, because improvising was never the job. The job was to follow the binder. So the kitchen freezes. And a freeze in the middle of service is a catastrophe.

That binder kitchen is the original way people tried to make machines intelligent, and understanding it is the fastest route to understanding what changed. For a long stretch of this field's history, making a machine seem smart meant what the binder suggests: a person sat down and wrote out the rules, by hand, one after another, trying to anticipate every case in advance. The technical name is a rule-based system. The most ambitious versions were called expert systems, because the aim was to capture a human expert's knowledge as a vast list of if-this-then-that instructions. If the patient shows this symptom and not that one, suggest this. If the chess position has this shape, prefer that move. The machine learned nothing. It executed a rulebook a person had written, quickly and without fatigue.

For tidy, bounded problems this worked, and in places it still does. The binder is too easily written off as a dead end. When the whole world of a problem fits in a binder a team can actually write, a rule-based system is fast, predictable, and, most important, inspectable. You can open the binder and read why the machine did what it did. That virtue is not small. We are going to miss it badly in later chapters, when we meet kitchens whose reasoning no one can read at all. If a rule-based system makes a mistake, you find the offending rule and fix it, and that ability becomes precious by contrast.

The trouble is that the interesting problems do not fit in any binder. Consider what it would take to write the rules for understanding ordinary language. Not the dictionary. Every way a sentence can bend. Take seven words: "He says my soup needs more salt." Now read it again, stressing a different word each time. Stress "he" and the verdict belongs to someone else. Stress "my" and it is my soup singled out, not the others on the pass. Stress "salt" and the fix is salt, not heat or time. One short sentence, seven readings, and nothing on the page to tell them apart. Now try to write the rules that capture that, for every sentence, in every context, with every shade of sarcasm and implication and regional habit. People tried. They wrote rules for years. The binder kept growing, and the kitchen kept hitting tickets it had no rule for, because real language, like real cooking, throws something new at you roughly every other minute.

Language is the vivid case. Every messy, real-world task ran into the same trouble. Teaching a machine to recognize a cat in a photograph makes the point general. You might start writing rules: a cat has pointed ears, so look for two triangles near the top. But now the cat is turned away. Or curled into a ball. Or half in shadow, or a breed with folded ears, or shot from above, or sitting beside a second cat, or next to a dog with suspiciously triangular ears. Every rule you write spawns a dozen exceptions, and every exception spawns its own. The binder for "what a cat looks like" is not just long. It is endless, and it never quite works, because the visual world, like language, is too varied for any handwritten list to contain. The rule-based approach did not fail because the people were not clever. It failed because the problems are too big for a binder.

The breakthrough was to abandon the binder, and the idea is strange enough to state bluntly. You stop writing rules altogether. You show the kitchen a staggering number of examples and let it work out the patterns for itself. You do not tell it the rule for a finished loaf. You show it a million loaves, done and underdone and burned, until it develops a feel for doneness that no one ever put into words. You do not write the rules for recognizing a cat. You show it an enormous pile of pictures labeled "cat" and "not cat," and let it learn cat-ness on its own. That shift, from a kitchen that follows written rules to a kitchen that learns from examples, is the entire foundation of what we now call AI. The technical name for the new way is machine learning, and the next several chapters are the story of how that learning actually happens.

Smart, skeptical people often say of modern AI, dismissively, that it is "just a bunch of rules" or "just a big if-then statement someone programmed." After this chapter you can see why that is wrong, and the wrongness is the whole point. Nobody wrote the rules for a modern learned system. No human sat down and specified how it should respond to your particular sentence, or what makes a photo a cat. It developed its own internal way of working by grinding through examples, and the result is something no one can read back as a list of rules. It is not a binder. It is a feel developed from examples, and that is why it can handle the strange ticket that would have frozen the old kitchen, and why, when it goes wrong, you cannot open it and find the bad rule.

The error we have just named, assuming today's systems are still, secretly, a giant binder, traps half the people who think about this. They are not a binder, and believing they are sets you up to expect a consistency they do not have and a readable list of reasons they cannot give. The other half fall for its quieter twin: the assumption that the binder is dead, that rules have been replaced. They have not. Hand-written rules still live all through these systems, wrapped around the learned part like the plumbing and the posted policies around a kitchen. The list of orders the kitchen will not cook. The hard limits on what it may say. The connective code that carries an order from the dining room to the cook and back. Much of that is still rules a person wrote on purpose. The modern reality is not learning instead of rules. It is a learned cook working inside a building that still runs partly on hand-written rules, and confusing the two will muddle everything when we reach the kitchen's house rules and hard limits.

The binder is still in service. The automation that sends you a shipping notice when a box is scanned is a binder: a rule fired. The filter that flags an email containing certain words is mostly a binder. And the guardrails around a learned kitchen are partly a binder too, a posted list of rules wrapped around a cook who learned. Rule-based systems did not lose. They found their place, doing the bounded, inspectable jobs they were always good at, often in partnership with the learned kitchen rather than in competition with it. The story is not that the binder was wrong. It is that the binder could not stretch to language and vision and the messy tasks, and learning from examples could.

A real binder, though, has a virtue the learned kitchen does not. A binder is readable. Any cook can open it and find the rule, and so can you. The learned kitchen has nothing like that inside it. In place of written rules it holds billions, now often hundreds of billions, of numbers, tuned during training, that together add up to its behavior, and no one can open that and read it the way you read a recipe. We will spend the mains, the long middle stretch of this book, figuring out where those numbers come from.

HOW THE FEEL FORMS

On her first morning, the new cook is handed one job: pull the bread when it is done. No one gives her a number. There is no chart on the wall, no target for the middle of the loaf, no forty minutes, no rule to copy onto her hand. The head cook points at the oven, says "you'll get it," and walks off. It is the only way anyone has ever learned to do the job.

She does not get it at first. The early loaves come out pale and damp in the middle. A few come out too dark. So she stands at that oven through the whole shift, pulling loaf after loaf, pressing the tops with two fingers, listening to the crust tick as it cools, breaking open the failures to see what went wrong. By the end of the week something has changed in her, though she could not tell you what. She walks past the oven, glances, and pulls a loaf without breaking stride. Ask her the rule and she shrugs. There is no rule she can say out loud. She just knows done bread now, the way she knows her own name.

That shrug is the hinge of this whole book, and the thing most people get wrong about these systems. She learned a skill no one could write down, by working through hundreds of loaves until the pattern lived in her hands. The binder tried, with its shade of brown, and that rule holds only until the loaf is strange; her feel is what the rule was standing in for.

She was not programmed. She was trained.

That is the same line the binder drew, seen from the other side: rules a person wrote, against a feel the kitchen formed. Understand it and the rest of this book follows. Miss it and you will misread these systems at every turn.

The plain name for all this is learning from data. You do not tell the system the rule. For the hard tasks there is no rule simple enough to write. Instead you show it an enormous number of examples and let it build its own sense of the pattern, the way the cook built her sense of doneness. What she did with loaves, these systems do with text, or images, or sound. The produce differs. The move is the same: see enough examples, and a feel for the pattern forms.

The task behind most of the chat tools you have used is humbler and stranger than the marketing suggests. At its core, a language kitchen is trained to do one modest thing. Given a stretch of text, predict what comes next. That is the whole of the basic training task. Show it the start of a sentence, "the cook reached across the counter for the," and have it predict the following word. Pan, perhaps. Salt. Towel. It makes a prediction. Training then reveals what the next word actually was, and the system nudges itself a hair toward being more right next time. Now run that loop not once but more times than you can picture, across a quantity of text no human could read in a thousand lifetimes.

Now the remarkable part. Get good enough at predicting the next word, across enough kinds of writing, and you are forced to learn almost everything else along the way. To predict the next word in a recipe, the kitchen must absorb how recipes flow. To predict the next word in a story, it must track who is in the room and what just happened. To finish "the boiling point of water at sea level is," it must have soaked up which words reliably follow that phrase. No one taught it cooking or storytelling or physics as subjects. It learned a feel for what text looks like, deeply enough that those subjects arrived with it, the way the cook, learning doneness, also picked up the smell of a too-hot oven and the sound of a good crust. The capability is a side effect of relentless prediction. That it works at all is one of the real surprises of the field.

One consequence follows. Ask one of these systems for the capital of some country and it will usually answer right, and it is tempting to conclude it "looked up" a fact. It did not. During training, across its produce, the words naming that capital kept turning up alongside the words naming the country, in sentence after sentence and context after context, so the kitchen's feel makes those words the likely completion. It is not consulting a fact. It is assembling the most probable next words, and for well-documented questions the most probable next words happen to be the true ones. That is why it is reliable on the common and the well-attested, and shakier where its produce was thin.

Now one of the most expensive misunderstandings people carry, and I will be plain about it. Learning a feel is not the same as understanding. The cook can pull perfect bread for forty years and never explain the chemistry of why heat sets a crust. Her hands know. She does not, not the way a food scientist knows. The kitchen is the same, and more so. It has a powerful feel for what good answers look like. Whether anything in a feel that large deserves the word understanding is a question the people who build these systems still argue; what no one disputes is what the feel lacks: no reliable check of its answers against the world, no reliable inner voice that says wait, that cannot be right. It has a feel, and a feel can be confidently, fluently wrong.

You see the gap most clearly when the feel gets fooled. Ask the kitchen a question that sounds normal but hides a false premise, "which of the two moons of our planet is larger," and a system running on feel alone may play along and compare the two moons, because the shape of the question was familiar even though its substance was rotten. The cook who learned doneness by feel can be fooled the same way, by a loaf that looks right on the outside and is raw within. Feel is fast and powerful and it is not knowing. Nothing in the kitchen catches the difference on its own.

The cook's feel lives in a person who also has a whole life and a fund of common sense to fall back on when the bread is strange. The kitchen has none of that. Its feel is all it has. There is no person back there to catch the odd case with a lifetime of judgment. That makes the feel both more impressive, because it built so much from examples alone, and more brittle, because when it fails there is nothing underneath to break the fall.

There is a quieter consequence too, and it bridges to the mains ahead. If the kitchen learns its feel from examples, then the examples are everything. A cook trained only on white sandwich loaves develops a beautiful feel for white sandwich loaves, a rougher feel for the loaves near them, and not much at all for a sourdough she has never seen. A learned feel stretches a little way past its examples, which is how the strange ticket gets handled, and it thins the further a request strays from everything it formed on. Whatever is in the produce becomes the cook's sense of the world. Whatever is missing or wrong becomes a blind spot she cannot even perceive. When the mains begin, we will walk into the storeroom and examine the produce, because once you accept that taste is learned and not programmed, only one question matters, the one that decides everything downstream: learned from what?

KITCHENS THAT COOK, KITCHENS THAT SORT

A KITCHEN FOR EVERY CUISINE

On a street where every door is a kitchen, no two make the same kind of thing. The first turns words into words: slide a question through the window and a written answer comes back. Next door, a kitchen turns words into pictures, taking a description and plating a painting of it. Across the way, one takes the sound of a voice and writes down what was said. Another runs that in reverse, reading text aloud in a voice so smooth you would swear a person was speaking. At the end of the block sits a quiet place that makes nothing you eat; it watches what you have ordered before and slides you a card suggesting what you will probably want next. Every one of these is built the way we have been describing, each having learned a feel from a mountain of examples, and yet each takes in a different kind of order and plates a different thing. "AI" is not one kitchen. It is a whole street of them, and treating them as interchangeable is a beginner's error.

The technical word for what a kitchen takes in and puts out is modality. Text is a modality. Images are a modality. Audio is a modality. Builders make a model for particular modalities: text in and text out, or text in and image out. When you hear a system called multimodal, strip the jargon and it means a kitchen that works in more than one modality, on the way in, on the way out, or both: a place might take in a photo with a question typed under it and tell you what the photo shows, or take plain text and answer out loud.

Work one of these through, the words-to-pictures kitchen. How does a kitchen learn to plate a picture from a sentence it has never seen? The same way the bread cook learned, by examples, except the examples are images paired with descriptions of what they contain. Show a kitchen an enormous number of pictures, each with words attached, and it gradually develops a feel for how descriptions relate to how things look. After enough of that, hand it a brand new description, words never paired with any image in its training, and it can cook up a matching picture, because what it learned was the relationship between words and appearance, not a fixed catalog of pictures to copy. The learning is the same trick as the bread, run on different produce. The cooking is not: a words kitchen lays its answer down in order, one piece after another, while a picture kitchen usually works the whole canvas at once, starting from something like static and refining it toward the description over many passes. Same lesson about learning from examples, a different motion at the stove, and worth holding onto when you meet the token-by-token account later, which is about language kitchens only.

Another example is the one quietly shaping your days, and almost no one pictures it as AI: the recommendation kitchen. It learned from an immense record of what people chose after what: watch this, and people tended to watch that next. Its produce is behavior, its output a prediction of what you will want, and it is built on the same learning-from-examples engine as the chatty kitchen down the street, even though it never says a word.

This explains a piece of news that confuses many people. You hear that a single system can now see and hear and speak, and it sounds like one mind waking up to the world. It is closer to a kitchen trained on several kinds of produce at once, so it can take orders and serve answers in more than one of them. Useful, impressive, and the same underlying move. No new magic arrives when a kitchen goes multimodal. Just more variety in the produce, and a kitchen that learned to work across it.

The beginner's error turns costly here: assuming that what a kitchen learned from one kind of produce carries over to another kind. It does not carry over on its own, because a feel comes only from the examples it was shown; a kitchen that handles both was trained on both. A kitchen with an astonishing feel for language did not thereby acquire a feel for images, any more than a master baker is automatically a master of the grill. When a company sells you one product that does words and pictures and sound, it is often several specialized capabilities behind a single window, or one system trained across several modalities but stronger in some than others. Excellence in one cuisine does not transfer to the rest for free, and you must judge each on its own plate rather than on the reputation of the others.

The second error is subtler. Even when a kitchen is multimodal, learning across modalities is not understanding the connection the way you do. When you see a photograph of a hot stove and flinch, that flinch rests on a lifetime of living in a body in the world. When a multimodal kitchen relates the word "hot" to the look of a glowing burner, it learned that relationship from examples of words paired with images, and nothing more. The crossover is real and it is still only pattern learned from produce. Keep that straight, because the more senses a kitchen appears to have, the stronger the pull to imagine a person back there, and we have already established there is no one.

In the real world, cuisines are cultural; they come from places and people and histories, and a chef choosing one is making a human choice. On the street of AI kitchens, whoever built the kitchen picked the cuisine, a kind of produce and a kind of plate, because that was the problem they wanted solved. There is no culture in the choosing. When the metaphor starts to make these systems feel warm and rooted and human, let the seam show: what makes one of them a different cuisine from the next is a different pile of examples, a different shape of plate, and in some cases a different way of working the stove.

THE KITCHEN THAT SORTS

Not every kitchen cooks. In the receiving area behind a large operation, one station creates nothing at all. Its only job is to sort. A worker takes each item off the incoming truck and drops it into the right bin: ripe here, unripe there, premium in this crate, seconds in that one, spoiled into the bin by the door. Nothing new is made. Every item that arrives gets a label and a destination. It is humble work, and the entire operation depends on it.

The sorting station comes second here, and not because it matters less. The kitchens that get all the attention, the ones that write and paint and talk, have given most people a lopsided picture of what AI is. They imagine AI as a thing that generates, that produces new content on demand. Much of the AI running the world does nothing of the kind. It sorts. It takes each input and assigns it a label or a score, and that dull work is among the most widespread and important uses of the technology. The technical names are classification, when the answer is a label, and regression, when it is a number, and the plain version is the sorting station: a kitchen that labels rather than creates.

You meet it constantly, mostly without noticing. The filter that decides which messages are junk and which reach your inbox is a sorting kitchen: not a word list fired by rule, but a feel learned from examples, dropping each message into the spam bin or the keep bin. The system that flags a transaction as possibly fraudulent is sorting: this one looks normal, that one looks suspicious. The tool that reads medical scans and flags the urgent ones for a doctor is sorting. The feed that ranks what you see is, underneath, sorting and scoring at a vast scale. So is the recommendation kitchen: the card it slides you is the top of a scored list. None of these generate a new plate. They look at what arrives and assign it a label or a number, and much of the value and the trouble of AI lives in this category.

The engine is the one you already understand. A sorting kitchen learns from examples, as the bread cook did, except the output is a label rather than a fresh creation. To build a spam sorter, you show the kitchen an enormous pile of messages already marked spam or not spam, and it develops a feel for what spam looks like, until it can label new messages it has never seen. Same loop of learning from examples, same feel forming from produce. Only the shape of the output is different: a bin, not a dish. Generation makes something new; classification sorts something given. Both are the same kind of learned kitchen wearing different aprons.

Why insist on this distinction in a book mostly about the generating kind? Because the sorting kitchens are often the more dependable of the two. A sorting kitchen has a narrower job, spam or not spam, fraud or not fraud, and a narrow job is easier to do reliably and easier to check, since you can count how often it sorted correctly. The generating kitchens get the headlines because they are flashier and more general, but generality is what makes them harder to trust.

Now the catch. A sorting kitchen can be confidently wrong, the same as any other, and a misplaced label has real consequences: the important message sorted into junk, the legitimate transaction flagged and frozen, the diseased scan waved through.

Stand in one of those for a moment. A card is refused at a counter with a queue behind it, because a purchase in an unfamiliar place resembled the pattern of a stolen card closely enough. Nothing was stolen. The sorter was working exactly as built, and none of that helps the person at the counter, who cannot argue with a bin. Worse, sorting kitchens inherit bias from their labeled examples as generating ones do, and less visibly, because a label feels objective in a way a paragraph does not. If the examples used to teach a sorter carried a skew, the sorter will sort with that skew while presenting each decision as a neutral fact. A score on a person, a flag on an application, a ranking of a candidate: each arrives looking like a plain fact, and a later chapter shows what is underneath.

One spot changes how you should read these systems. A real sorting station drops each item into one clean bin. A sorting kitchen usually does something softer: it produces a probability or a score, this message is eighty percent likely to be spam, and then someone, or some rule, draws the line at which a score becomes a bin. That means two things. The line is a choice, and moving it trades one kind of error for another, more spam caught at the cost of more good mail lost. And the crisp label you finally see, spam, fraud, approved, hides an underlying uncertainty that was rounded off somewhere upstream. When a system hands you a clean category, remember there was almost always a wobblier number behind it, and a human decision about where to cut.

WHY IT GOT SO GOOD SO FAST

ONE DISH, NOT EVERY DISH

People cross a city for this kitchen, and it makes exactly one thing. A single dish, perfected over years, so good that a line forms before the doors open. If you go, you order that dish. The mistake, which visitors make constantly, is to assume that a kitchen this good at one thing must be good at everything. So someone orders off-menu, asks for a dish the place has never specialized in, and what arrives is fine. Edible. Forgettable. The same hands that make the famous dish produce a mediocre version of the thing actually requested, and the diner sits there confused, because how can a kitchen this celebrated be this ordinary.

Because being great at one dish and being great at every dish are different things, and the gap between them is where expectations of AI go wrong, in both directions at once.

The plain idea is the difference between narrow ability and general ability. A narrow kitchen is shaped, by its training and design, to do a particular range of things well. A general kitchen would handle anything thrown at it with equal skill. Almost everything in use today sits far closer to the narrow end than the breathless coverage suggests, even the systems that feel as though they can discuss anything. They feel general because language touches everything, so a kitchen with a strong feel for language can hold a conversation about cooking, law, grief, and tax, and sound at home in all of them. But sounding at home in a conversation about a thing is not the same as being reliably skilled at the thing, and the difference surfaces the instant the task stops being "talk about it" and becomes "get it exactly right."

The worked example surprises people every time. Take a strong language kitchen and ask it to turn a long, tangled email thread into a clear summary anyone could act on. It will do this beautifully. Now ask it to multiply two long numbers and give the exact result. It may well return a confident, wrong answer, off by a digit somewhere in the middle. How can the same system write like that and fumble arithmetic a child's calculator nails every time? Because the kitchen never learned to calculate. It learned the feel of language, and "what a multiplication answer looks like" is part of that feel, so it produces a number-shaped answer that is often close and sometimes wrong. It is cooking the look of a correct answer, not computing one. The house dish is language. Arithmetic is off-menu, and it shows.

The lesson generalizes past arithmetic. Ask the kitchen about something that happened last week, and a plain language kitchen cannot know, because its feel froze when its produce was gathered. It may still turn out a confident, specific-sounding answer about the recent event, because answering questions is the shape it knows, and that answer can be invented. Two different off-menu orders, arithmetic and current events, and one underlying truth: the kitchen plates what it is bad at. It plates a confident version anyway. We will see later how kitchens patch both of these by sending out to a calculator and a search runner, but the bare system, on its own, fakes the off-menu order rather than declining it.

You will meet people stuck on each side of this. On one side stand those who underestimate, who hear "it is basically autocomplete" and conclude these systems are toys. They are not toys. A kitchen with a deep feel for language is powerful and broadly useful, because so much of human work runs on language: writing, summarizing, explaining, drafting, reorganizing, translating between ways of saying things. Dismissing that as "just autocomplete" is like dismissing a master baker as "just someone who heats dough." Technically adjacent to true, and missing everything that matters.

On the other side stand those who overestimate, who see one dazzling plate and conclude the kitchen can do anything. They hand it a task that needs real calculation, or current information it never had, or a long chain of careful reasoning where one wrong step poisons the rest, and they trust the confident plate that returns because the last one was so good. This is the more dangerous error, because the kitchen's fluency promises what its reliability cannot always deliver. The famous dish lulls you into trusting the off-menu order, and the off-menu order is where it is weakest and where it will still sound certain.

Calibration is the skill: not "AI is amazing" and not "AI is overhyped," but a working sense of which orders this kitchen is good at and which it merely sounds good at. That sense is learnable, and here is a practical way to develop it. Before trusting an answer, ask three quick questions. Is this the house dish, the language-shaped work the kitchen excels at, or is it off-menu: exact calculation, current facts, long multistep reasoning where it tends to fail? How common was this in its likely produce, well-documented and central, or niche and thin? And can I cheaply verify it if it matters? Those three questions, asked in seconds, will steer you right far more often than either blanket trust or blanket suspicion. And when they leave you unsure, one blunt test almost never fails: run a small piece of your own real work through it and check the result. Hold onto that habit; it grows into the way you grade kitchens later in the book.

One place the picture flatters the real thing, and the flattery is the danger itself. A real one-dish kitchen knows its limits. Even the cooks who plate your off-menu order without a word know it is not their strength and will say so if you ask. The AI kitchen has no such self-knowledge. It does not know which of your orders are its house dish and which are off-menu. It plates the arithmetic with the same confidence as the summary, and it cannot be trusted to warn you that it has stepped outside what it is good at. That missing self-awareness is why the judgment must live with you, the diner, and never with the kitchen.

MOSTLY A MATTER OF SCALE

For years this kitchen was unremarkable. It was on the same corner, with the same basic idea it has now, and the food was middling. Regulars were polite about it. Then, over a stretch that felt sudden from the street, it became extraordinary, the sort of place written about and imitated. People assumed the owner had discovered some secret technique, a new method, a stroke of genius in the back. The truth was less romantic. The owner did the same thing, much bigger. A far larger market to buy from. A vast bank of new stoves. And a far bigger room to hold it all. The leap was mostly a matter of scale.

That story is, at bottom, the story of why AI got so good so fast, and it corrects two opposite myths: that there was a single magic breakthrough, and that the improvement was inexplicable. Neither is right. The dominant driver of the leap was scale, growth in three things at roughly the same time, and the plain version is the whole operation getting bigger in every dimension at once.

You have already met all three. The first is data, the produce: far more examples to learn from than before, gathered at a scale earlier kitchens never had. The second is compute, the stoves: enough raw cooking power, from specialized hardware, to run the enormous training loop on all that produce in a reasonable time. The third is model size, the number of dials, the internal settings that training slowly turns: a kitchen with far more of them, capable of holding a richer feel. Pile up more produce, more stoves, and more dials together, and the kitchen's ability climbed, again and again, well past what anyone initially expected. A design change also made bigness practical, the arrangement we will meet in the chapter on how every word weighs on what came before. But the engine of the leap was scale.

The surprising part, the thing that makes this more than "bigger is better," is how the improvements arrived. They did not always come smoothly, a little more size buying a little more skill. Sometimes a kitchen would be unable to do a thing at all at one size, and then, given more dials and fed more, it could suddenly do that thing, as if an ability had switched on. Capabilities appeared that no one had explicitly built in, seeming to emerge from scale alone. This is part of why the field moved so fast and why it kept surprising even the people inside it. One note, and it matters more than a footnote. Researchers genuinely argue about how real these sudden "emergences" are, and good evidence suggests that many of the apparent overnight jumps are an artifact of how the test scores a capability: a smooth underlying gain that the test, grading pass-or-fail, renders as a cliff. Treat the dramatic "it suddenly could do X" stories with caution. Either way, the practical lesson holds, because you still could not always look at a smaller kitchen and predict the bigger one. You had to build it and find out.

That unpredictability has a direct and humbling consequence, the first of three things worth taking from the scale story. Whether the gains arrived as real cliffs or as smooth curves the tests drew as cliffs, almost nobody called them in advance, so anyone forecasting the future of these systems with confidence, in either direction, is guessing. The history of this field is full of confident predictions about what these kitchens would never be able to do, made shortly before a bigger kitchen did exactly that. It carries just as many breathless predictions of imminent miracles that did not arrive on schedule. Treat both kinds of certainty with suspicion. Scale has produced repeated surprises and we should expect to keep being surprised. Dressing a guess as a forecast is a mistake.

The second thing to drop is the assumption that bigger always wins and will keep winning forever. It will not, for reasons the cost chapter will make concrete. Each doubling of the kitchen buys a smaller improvement than the doubling before it, and that has held on average even through the surprises. Meanwhile the stoves and the produce and the dials all cost money, and those bills climb steeply, so the gap between what you spend and what you get back keeps widening. Scale drove the leap, and scale is not a free or infinite lever, so the field increasingly chases efficiency, getting more from a given size, rather than only chasing raw bigness.

The third matters most for your judgment. Making the whole operation bigger made it a better cook. It did not make it a different kind of thing. A larger kitchen still learns a feel rather than understanding, still produces plausible plates, still has no inner sense of what it does not know. Scale turned up the capability dramatically and left the fundamental nature intact, and with it the repeatable ways of going wrong. A bigger kitchen's mistakes tend to come out more convincing, even where they come out less often. Do not mistake the dramatic rise in skill for a change in kind. It is the same kind of kitchen we have been describing, run much larger.

The owner's bigger market, new stoves, and larger room are easy to picture; the real version is data, specialized hardware, and dials at a magnitude hard to imagine, concentrated among the few who can afford it. But the shape is right: the leap came less from a single secret than from doing the same thing, much bigger.

THE PRODUCE DECIDES EVERYTHING

The most important moment in a kitchen's day happens before a single customer arrives, at the back door, in the gray light of early morning. The deliveries come in. Crates of vegetables, boxes of fish on ice, sacks of flour. A cook who knows her trade does not sign for them and walk away. She opens the crates. She smells the fish. She turns a tomato in her hand, presses it with her thumb, and sends back the lettuce gone slimy at the edges. Everything that will reach a plate tonight is already shaped here, by what came through that door. A brilliant cook with bad produce makes bad food. No technique on earth turns a spoiled fish into a good dinner.

That is the truest thing about these systems, and it is routinely overlooked. Four chapters have established that the kitchen learns its whole feel from examples. This chapter is about the examples, and about what happens to your words on the way in. The examples are the produce, and the produce decides nearly everything.

The technical name is training data: the body of examples a system learns from. For a language kitchen, an ocean of text. For an image kitchen, a vast pile of pictures, usually with descriptions attached. When people say a model was "trained on the internet," they mean the builders gathered the produce at tremendous scale from text and images already out there. The amount is hard to picture. This is not a library. It is a stack of reading no working life could dent, run through the kitchen until a feel for it formed.

Two things about produce matter, and people fixate on the first while forgetting the second. The first is quantity. More examples generally means a richer feel, the way a cook who has handled ten thousand fish reads a fillet faster and truer than one who has handled ten. Much of the leap in the last chapter came from more. But quantity is the easy half. The half that decides whether you can trust the kitchen is quality, and quality is where it gets uncomfortable.

Two kitchens are alike in everything except what the builders fed them. For the first, the people who gathered the produce chose it carefully: accurate, varied, drawn from many kinds of writing and many points of view, checked and sorted. For the second they took whatever was cheapest to grab in bulk, mistakes and all, heavy on a few loud kinds of writing and thin on the rest. Both will cook with confidence. Both will plate fluent, professional-looking food. But the second learned its feel from bruised produce, and that taint runs through everything it makes, in ways you cannot see from the dining room.

This is the moment for a phrase you know: garbage in, garbage out. Whatever is wrong in the produce becomes wrong in the cook's feel. If the examples carry a common mistake, the kitchen learns the mistake as though it were correct, because it has no way to know otherwise. It never checked the produce against the world. The produce is its world. If the examples lean toward one way of seeing things, the kitchen's feel leans the same way, and it plates that lean as though it were plain fact. A later chapter takes this up, a kitchen that only ever learned one cuisine. Its root is here, in the crates at the back door.

A second consequence of learning from gathered produce explains a behavior that puzzles people. Whoever assembled the produce stopped gathering it at some point in time, training happened, and the dials froze. From that moment, the kitchen's feel is fixed around the world as it stood when they stopped. It does not know what came after. Ask about something recent, past the point where the gathering stopped, and it cannot know, because that event was never in its examples. The term you will hear is a knowledge cutoff. No mystery to it now. The kitchen learned from a market that closed on a certain date and has cooked from that memory ever since, no matter how much later you walk in. Anything fresher than the storeroom must be carried in from outside, which is the runner's job in a later chapter.

What can go wrong here hides where most people never look: the supplier list. In a real kitchen you can, in principle, ask where the fish came from. With most of these systems you cannot. The builders often do not disclose the exact contents of the training data, so you frequently cannot tell what they fed the kitchen, what they overrepresented, what they left out, or what mistakes they let through. You are tasting the food with no sourcing card. It means a confident answer is not a sourced answer, and the confidence tells you nothing about whether the produce behind it was good. When the stakes are real, that gap is where you bring your own judgment, or check the plate against a source you trust.

A subtler burn sits beside it. People assume that because the builders fed a kitchen so much produce, they must have fed it everything, evenly. They did not. Whatever kinds of writing and images were plentiful and easy to gather are overrepresented. Whatever was rare, private, in a less common language, or not online in bulk is thin or absent. The feel is richest where the produce was thickest and patchy where it was thin. So the kitchen can dazzle on a common, well-documented topic and mislead on a niche one, sounding just as confident in both. The evenness you imagine in those crates is not there.

Produce in a real kitchen is physical and perishable. You use the tomato once and it is gone, and freshness means rot. Data is not like that. It can be copied without end, used to train many kitchens, and it does not spoil on a shelf. What maps to freshness is not decay but recency: how up to date the examples are relative to the world you are asking about. When this book speaks of fresh produce, hear "recent and relevant examples," not "did not go bad." The handling discipline matches a real kitchen's; the reason behind it does not.

MISE EN PLACE

In the quiet hours of the afternoon, long before service, the cooks do the least glamorous and most essential work of the day. Prep. Everything that will be cooked tonight is broken down now into standard pieces. The onions are diced to a uniform size, not for looks, but because pieces of the same size cook at the same rate, and a line moving fast cannot stop to babysit one fat chunk that is still raw while everything around it scorches. Professional kitchens have a phrase for this, borrowed from the French: mise en place. Everything in its place. The whole night depends on the chopping nobody sees.

Here is something that will surprise you and explain some of the strangest things these systems do. The kitchen does not read your words the way you do. Before it can work with your sentence at all, your text is chopped into standard pieces, and those pieces are not quite words. They have a name you will meet on bills, in limits, and throughout the technical talk: tokens.

A token is a standard small chunk of text, the unit the kitchen works in. Sometimes a token is a whole short word, like "cook" or "the." Sometimes it is a piece of a longer word, so that "unbelievable" might be chopped into "un," "believ," and "able," three tokens for one word. Sometimes a token is a space followed by a word, because in the flow of text the spaces carry information. The builders worked out the exact way of chopping from the produce itself, to break text into pieces common enough to be useful, and they did it once, separately, before the kitchen ever formed its palate, its trained sense of what comes next. The next chapter is all about that. Every kitchen built this way handles chunks rather than letters, and they all share the same clumsiness with individual letters. You do not need the machinery, only the picture: before your sentence enters the kitchen, it goes through prep and comes out as a string of standard chunks. Standard does not mean even. The cook's dice match in size; these chunks do not. What makes them standard is that every one is a piece prep already knows.

One example, walked through. Take "The soup was unbelievable." To you, that is four words and a period. To the kitchen, after prep, it might be the chunks "The," " soup," " was," " un," "believ," "able," and a final chunk for the period. Seven pieces where you saw four words, each carrying the space in front where there is one. The count came apart on the long word, which is the whole point. Now take an unusual word the kitchen rarely saw, a rare technical term or an invented name. That one gets no tidy single chunk. It shatters into several small pieces the prep station recognizes. How badly a word shatters tracks how often prep met that exact spelling in the produce. The list of chunks prep works from was built to a fixed size, so a spelling earned its own chunk by being common enough to claim a place on it; "the" did, and so did "cook." A word you know well can still miss the cut: in our example, "unbelievable," for all the times people write it, was not common enough, so it came through in three pieces prep knew well. A word prep barely met comes through in crumbs. The kitchen works with whatever prep handed it.

Once you hold this picture, a family of famously strange failures stops being strange. People love to catch these systems failing at things a child does easily: counting how many times a certain letter appears in a word, spelling something backward, picking out words that start with a given letter. The kitchen fumbles these more often than you would expect, and prep is the reason. It works in tokens, the pre-chopped chunks, not in individual letters. Asking it to count the letters in a word is like asking the line cook to work from the diced onion instead of slicing the vegetable herself: she can often tell you roughly what went in, but she is handling chunks, not cutting letter by letter. So anything that turns on the individual letters, counting them, reversing them, sorting words by them, is something the kitchen does indirectly and unreliably, not naturally. The letters are not truly gone; they are just not the unit the kitchen works in.

One caveat: a polished product may still answer these correctly. Sometimes the system hands the task to a separate step that counts the ordinary way, and sometimes the kitchen has learned a workaround of its own, slowing down to spell the word out piece by piece before counting, the way a cook can recite what went into the dice without ever having sliced it. Either way it is working around its prep, not seeing through it, and tasks that turn on exact letters stay less dependable than the work that fits its units.

The first time this picture paid for itself, I was reading about a mistake. Someone had caught one of these systems miscounting the letters in an ordinary word, and the failure was going around as proof that the whole business was hollow. It did not look hollow to me. I had recently learned what happens to text before it ever reaches the kitchen, and from inside that picture the mistake was not absurd, it was predictable: a system handling pre-cut chunks was being asked to do the one job that needs the uncut vegetable. It was not a sign of stupidity. It was a sign of prep.

That is what the picture buys you. From outside, the failure looks like a hole in the whole idea. From inside, it looks like a boundary you can see the shape of, work around, and stop being surprised by.

This also demystifies a practical thing you bump into constantly: why everything is counted in tokens. The price of using these systems and the size of what they can hold at once are both measured in tokens rather than words, because tokens are the units the kitchen handles. As a rough guide, a token runs a little shorter than a word, so a page of ordinary text is more tokens than it is words. When you read that a system has some large limit "in tokens," translate it the way you would any unfamiliar unit: a certain amount of text, measured in the kitchen's prep pieces rather than your words.

In a real kitchen, uniform prep is about physics: even pieces cook evenly. In our kitchen, the chopping is about needing one fixed list of pieces that can still spell anything. Whole words would make an endless list, since new names and new coinages never stop arriving, and single letters would make every order far too long to work with.

Two failures follow from prep. The first is the letter trouble you have just seen. The second, most people never hear about. It matters for fairness. The builders shaped the chopping mostly around the text that was common in the produce. Text in a widely represented language tends to chop into fewer, cleaner tokens; text in a less represented language, or a different writing system, often chops into many more pieces for the same meaning. Since cost and limits are measured in tokens, the same thought can literally cost more and take up more room in one language than another. A choice made deep in prep, for sensible reasons, lands unevenly out in the world.

A MILLION PRACTICE NIGHTS

TEACHING THE KITCHEN TO TASTE

A cook learning to season a sauce does it the same way in every kitchen on earth, and watching it is the closest you will come to seeing how these systems learn. She tastes the sauce. She holds it against the target in her head, the way it is supposed to taste. She decides it needs a little salt, and adds a few grains, not a fistful, a few. She stirs and tastes again. Closer. A touch more. Tastes again. She is not following a recipe that reads "add 4.2 grams of salt." She is running a loop: taste, compare to the target, make a small adjustment, taste again, around and around, narrowing in until the sauce sits right. Then she does it for the next sauce, and the next, for years, and somewhere in those years a palate forms that can walk up to any sauce and know what it needs.

That loop is training. Understand the sauce and you understand it, because it is, stripped to essentials, how one of these systems learns. The scale is inhuman and the thing being adjusted is strange, but the loop is the loop.

When a kitchen is brand new, before any training, it knows nothing; its internal settings are essentially random. Ask it to predict the next chunk and out comes nonsense, the equivalent of a cook with no palate dumping random handfuls into the pot. Now the training loop begins. You show it a real example from the produce, a stretch of text, and have it predict the next chunk. It predicts, badly at first. Then you reveal what the next chunk was. That is the target, the way the sauce is supposed to taste. No one had to write that target down for the lesson; it is the chunk that actually came next in the real text, so every passage in the produce is its own answer key, with no human labeling required, and the kitchen practices on many such passages at once rather than strictly one by one. The system measures how far off its prediction was, then makes a tiny adjustment to its internal settings, nudging them a hair in the direction that would have made the prediction less wrong. Then it does it again, with the next example, and again, across the whole mountain of produce, an unimaginable number of times.

The palate is the kitchen's trained feel, the same thing a cook means by the word, and the thing being adjusted to build it is a collection of internal numbers called parameters, or weights. The feel is what the kitchen has; the numbers are where it lives. Picture them as dials. Not a few, the way a stove has a few. Millions of them in a small kitchen, many billions in a large one. Each dial is just a number, and alone it means nothing. Together, their exact settings produce the kitchen's behavior, the way the precise tension in thousands of a cook's habits and instincts together produces her palate. When we say the system "learned," we mean this: the training loop slowly turned those billions of dials, a hair at a time, example by example, until their combined setting produced good predictions. Nobody turned the dials by hand. No person could; there are far too many, and no human knows what any single one should be. The loop turned them, automatically, through the relentless taste-compare-adjust grind.

One turn, concretely. The produce contains a sentence like the one you tried to finish a few chapters back: "The cook reached for the salt." Training feeds the kitchen "The cook reached for the" and asks for the next chunk. The untrained kitchen predicts "bicycle." The loop reveals the truth was "salt," measures that "bicycle" was wrong, and nudges the dials a hair so that next time, in a context like this, "salt" and its neighbors grow a little more likely and "bicycle" a little less. One nudge changes almost nothing. But run that loop across billions of examples, every kind of sentence, every topic, every turn of phrase, and the dials settle into a configuration that makes startlingly good predictions. The feel we met with the bread, that wordless sense of what comes next, simply is the settled state of all those dials. There is nothing else in there: no facts filed where anyone could look them up, no rules written down. What it knows is smeared across the dials, reachable only by cooking. Only dials, settled by the loop, that happen to turn out knowledgeable-seeming food.

You know from the opening chapter that the kitchen cooks its answers rather than fetching them. Here is the stranger fact underneath, the one that trips up even careful people: there is nothing in there to fetch. The trained kitchen is not a database. It did not file away the produce the way a library files books, and it cannot reliably pull up a given document and read it back, any more than the cook keeps every sauce she ever tasted in a freezer. The produce passed through the loop and left its mark in the setting of the dials, and most of it, as a stored thing, is gone. What remains is mostly a palate, not a freezer full of saved meals. This is why the kitchen can answer about things its exact examples never contained, the way the cook can season a sauce she has never made: the feel generalizes. It is also why it cannot reliably quote its sources. It mostly tasted the produce and kept the feel, not the documents.

One qualification, because it matters for a later chapter. A trained kitchen can, in fact, sometimes reproduce bits of its produce more or less verbatim, especially passages it saw many times over or rare, distinctive strings that stuck. It is not a database, and it is not entirely free of copies either. The accurate picture is a palate that occasionally remembers an exact taste. That makes two questions more tangled than "it kept no copies" suggests: what it learned from, and whether its plates can come too close to a specific source. Hold that; we return to it when we ask who owns what the kitchen makes.

The trouble starts in what the dials will not tell you. First, the dials are unreadable. No line in there reads "salt follows reaching." Instead there are billions of numbers whose individual values mean nothing to any human. When something goes wrong, you usually cannot open the kitchen, find the bad rule, and fix it, because there is no rule, only the diffuse setting of countless dials. This is the deep reason these systems are so hard to fully explain or correct. We traded the readable binder of the old kitchen for a palate no one can inspect.

Second, training is enormous and one-time, while using the kitchen is small and constant, and confusing the two costs people money and worry. Teaching the palate takes weeks or months for the largest kitchens, and enormous equipment and money, spent once, up front, before anyone orders. After that, cooking a single plate to order is comparatively quick and cheap. Bigger stoves and bigger bills come later, as the orders pile up, but hold this fixed: the expensive part already happened before you walked in. You are paying to have a plate cooked, not to have a palate trained.

Third, and this is the bridge to a chapter that surprises everyone, the dials freeze. When training ends, the settings lock, and that palate is the kitchen you use. It does not keep learning from your conversations. When you correct it, you are not turning its dials. Nothing changes because one diner sent a plate back. A later chapter takes up what that means, a kitchen with no memory, because it has real consequences for how much you can rely on it to "learn your preferences." For now, the headline: the palate was set during training and then frozen, and your chats happen afterward. It stays frozen until its makers train a fresh version and put it in place, which is a new kitchen rather than this one having learned.

A cook's palate lives in one head, and she can usually tell you, roughly, why a sauce needs salt. The kitchen's palate lives in billions of separate dials, and no one can read a reason out of them. Ask it why and you will get a fluent answer, cooked the same way as any other, which is not the same as a report from inside the machine. And a real cook keeps learning every shift, her palate always shifting a little. The trained kitchen's palate stops the moment training ends and never changes again on its own. When the kitchen feels like a cook getting to know you, hold on to the bend: the palate is frozen, and the sense of being known comes from elsewhere in the restaurant.

THE FOURTH DISH

Two cooks start out looking identical. Both are talented, both spend a year practicing in the same kitchen with the same equipment and the same patient training. The difference is what they practice on, and it will turn one into a cook you would hire anywhere and the other into a cook who falls apart the moment service gets real.

The first cook practices on variety. Every night, different dishes, different ingredients, different problems. Some nights the fish is gorgeous, some nights it is mediocre and she has to adapt. By the end of the year she handles whatever walks in the door, because whatever walks in, she has seen something like it and learned the underlying moves, not just the specific dishes. The second cook practices on three dishes. The same three, every night, for a year. He becomes flawless at those three, faster and cleaner than the first cook on those exact plates. Then a ticket comes in for a fourth dish, just outside his three, and he is lost. He never learned to cook. He learned three dishes by heart, and a fourth exposes that his practice went deep but not wide. He memorized. He did not learn.

That difference carries formal names; it is one of the central tensions in training these systems. The good outcome is generalization: the cook developed a feel that extends to new situations she never practiced on. The bad outcome is overfitting: the cook fit himself so tightly to his exact practice material that he can only do that, and stumbles on anything new. Every kitchen we use is trying to land on the first and avoid the second, and it is harder than it sounds.

Start with the practice itself. A kitchen often goes through its examples more than once, and each full pass has a name you will see: an epoch. A modest kitchen may practice the same material many nights running; the very largest, whose produce is too vast to be worth repeating, may get through the whole of it only about once, returning mostly to the best crates. With each pass the dials settle a little more, the predictions improve, and the people training the kitchen watch one number closely: how wrong it is on average, totaled up. That number is the loss. Training is, in large part, the long project of watching that number fall, night after night, as the palate sharpens.

Here is the trap, beautiful and counterintuitive. You might think the goal is to drive that loss as low as possible, to make the kitchen as right as it can be on its practice material. That is exactly how you get the second cook. Let a kitchen practice too long and too hard on a fixed, narrow set, and it stops learning the general feel and starts memorizing the specific examples. Its performance on the practice material keeps improving, the loss on those exact examples keeps dropping, and meanwhile its ability to handle anything new rots. It looks like improvement right up until you hand it a real order it did not practice on. The lowest loss on the practice plates is not the goal. The goal is a feel that travels.

So how do the people training these systems avoid raising the second cook? Mostly through two moves, both mapping cleanly to the kitchen. The first is variety in the produce, the reason the produce chapter mattered so much: a kitchen fed a wide, varied range is far harder to overfit than one fed a narrow range, as the first cook's varied practice protected her. The second is a kind of pop quiz. During training, you hold back some examples the kitchen never practices on, and you test it on those. If it does well on plates it never practiced, it is generalizing. If it aces its practice plates but flunks the held-back ones, you have caught it overfitting, and you stop or adjust before it worsens. It is the difference between testing a cook on his three drilled dishes and testing him on something off his list. Only the second test reveals whether he can cook.

This also answers a common question: is more always better? More produce, more practice, a bigger kitchen with more dials. For a long stretch, more meant better, and much of the dramatic improvement came from scaling up, as we saw. But "more is better" is not a law of nature, and it bends. Diminishing returns arrive here too, and more practice on the wrong, too-narrow produce does not help; it just overfits faster. The lesson is subtler than "more is better." The right practice, varied and broad, generalizes, and you can absolutely practice your way into being worse at the actual job.

For you, who will never train one of these, two practical consequences follow. First, a kitchen can look brilliant on a test and fall apart on real work, and overfitting is one reason. When you see a system score astonishingly well on some test, your first quiet question should be whether it generalized or had practiced on something too close to the test. A cook who has secretly seen the exam looks like a genius until the questions change. This is not an accusation; it is calibration. Dazzling performance on familiar material tells you less than solid performance on new material. Second, more day to day, these systems are strongest on the kinds of orders well represented and varied in their practice, and shakier on the unusual, the niche, the novel. When your request sits near the broad center of what the kitchen practiced, you get the first cook. When it is strange or far from anything common, you are more likely to find the edges, where the feel thins.

And as with the off-menu order before, the overfit kitchen cannot feel its own edge. The second cook at least knew he was lost; the kitchen does not. It plates the untrained fourth dish as surely as the three it drilled and never marks which is which. The practice happened; the sense of what the practice missed did not.

EVERY WORD WEIGHS ON WHAT CAME BEFORE

THE FLAVOR MAP

A serious cook does not organize the walk-in alphabetically. Alphabetical would put the anchovies next to the apples, which helps no one at six in the evening. She organizes by something more useful: she keeps things that behave alike near each other. The acids cluster together, lemon beside lime beside vinegar. The soft herbs sit in their own corner, basil near mint near cilantro. The deep, savory things gather on one shelf. Walk her walk-in and you are walking a map where nearness means similarity, where two things sitting close together taste or act alike, regardless of how their names are spelled. That map is one of her most powerful tools, because it lets her reach, without thinking, for the thing that will work in place of the thing she is out of.

These systems build a map much like it, one that powers much of what you do online. The technical name is embeddings, and the plain version is the flavor map: a way of placing every word, and every thing, as a point in a vast space where nearness means similarity of meaning. On this map, "happy" sits near "glad" and "cheerful," far from "furniture." "Doctor" sits near "nurse" and "hospital." The kitchen learned the whole arrangement from its produce, by noticing which words turn up in similar company, and the result is a map where meaning becomes distance, where related ideas are neighbors and unrelated ones are across town.

How does a map like that get built? The same way everything else here gets built, from examples. Words that appear in similar contexts across the produce, surrounded by similar neighbors, end up placed near each other, because the kitchen learned they play similar roles. No one drew the map by hand. It emerged from the produce, the way the cook's sense of which ingredients are interchangeable emerged from years of cooking, not from a chart someone handed her. And once you have such a map, you can do something useful that older methods could not: you can find things by meaning rather than by exact words.

Old-style search matched words. Search for "ways to cut costs" and it looked for documents containing those exact words, and missed a perfect document titled "reducing expenses," because not one word matched. Search built on the flavor map works differently. It places your query on the map and finds whatever sits nearby in meaning, so "ways to cut costs" lands right next to "reducing expenses," and you find the document you actually wanted even though it shared no words with your search. This is called semantic search, search by meaning, and it is why modern search, recommendations, and the pantry-checking a few chapters ahead all work as well as they do. When a system finds the relevant thing despite different wording, the flavor map is usually underneath.

It is how a recommendation kitchen knows that someone who liked one thing might like another that seems unrelated by name but sits nearby in behavior. It is how the retrieval that grounds a good answer, coming up soon, decides which pantry documents are relevant to your question. It is what search and matching across modern systems run on. You rarely see it, and you rely on it constantly.

The map was built from produce, like everything else here, so it inherits that produce's skews. If the produce placed two things near each other that should not be near, by reflecting a stereotype rather than a truth, the map encodes that, and every search and recommendation built on it quietly carries the skew. A flavor map learned from biased produce will, for instance, place certain jobs nearer certain kinds of people, not because that is true, but because the produce talked that way. And because the map feels mathematical and objective, that encoded skew can be even harder to notice than a biased paragraph. "Near on the map" is a learned statistical fact about the produce, not a verified truth about the world.

The walk-in picture fails you here, and the failure matters. A real walk-in is a room you can walk, with shelves you can see, in three dimensions. The flavor map has nothing so comfortable. It lives in a space of hundreds or thousands of dimensions, far more than anyone can picture, because meanings can be alike in many ways at once, in topic, in tone, in role, and a map with only a few directions cannot keep all those ways separate. Nearness on it is real and computable and impossible to visualize. So keep the walk-in for the intuition, near means similar, but know that the actual map has more dimensions than you can imagine. That is what lets it capture so much, and why no one can simply look at it.

THE BRIGADE

A single plate does not get made the way you probably imagine. You might picture one cook taking your order and producing the whole dish, like a painter at a canvas. Serious kitchens do not work that way. Your order moves down a line of stations, and at each station one cook does one thing and passes it on. The first handles the raw protein. The next adds the sauce that reduced all afternoon. The next plates it, the next garnishes, the next wipes the rim and sets it in the window. No single cook made your dish. The line made it, each pair of hands doing something simple, the dish growing more finished as it travels.

That line has a name in professional kitchens, again from the French: the brigade. And it is the best picture I know for the thing at the center of these systems, the thing you have heard called a neural network. A neural network is, in plain terms, a line of simple stations an order passes through, each transforming it a little, until a finished result emerges at the far end.

No equation is needed to define it. A neural network is built from layers. A layer is a station on the line. Your input, the tokenized order from the last chapters, enters at the first layer. That layer does something simple and mechanical to it and passes the result to the next. That layer does its own simple operation and passes it on again. The order flows through layer after layer, each a station, each transforming what it received, and out the far end comes the kitchen's response. When you hear "deep learning," translate it: deep means a lot of stations. A deep network is a long line. The depth that sounds mysterious is only the number of stations the order passes through.

The natural question is how a chain of simple stations produces anything as impressive as a fluent answer. The almost unsatisfying truth is the one the whole field rests on: chain enough simple operations together and tune them all with training, and the combination captures astonishingly complex patterns, even though no single station does anything you would call clever. The early stations tend to pick up simple, local features of the input. Later stations combine those into bigger, more abstract patterns. In an image kitchen, the first stations might react to edges and patches of color, the middle ones to shapes, the late ones to whole objects, though that tidy ladder is the picture for a kitchen that sees; in one that reads, what each station settles on is murkier and rarely so cleanly nameable. No one assigned those jobs. Training set the dials at every station so that, working together, the line as a whole does the task. The intelligence, such as it is, lives in the chain, not in any one link.

What flows down the line is not your literal words but a running, evolving set of notes about your order that grows richer at each station. The notes do not start blank: each token of your order enters the line as its point on the flavor map. The first station jots something basic. The second reads those notes and adds something of its own. By the time the notes reach the last station, they have been enriched, step by step, into something detailed enough to produce a good response. Each station only ever does its small job on the notes it received. The finished plate is the accumulation of all those small jobs in sequence. That is a neural network passing information through its layers, with none of the math that usually scares people off.

The first mistake is imagining a little thinker at each station, a tiny chef making decisions. There is none. Each station performs a fixed, simple operation, the same every time, with its own dial settings. The richness comes from the chain and the training, not from cleverness at any one station. When you catch yourself picturing someone in there making choices, let it go. It is stations doing simple operations in sequence, fast.

The opposite mistake is hearing "just simple operations chained together" and concluding the whole thing must be simple or fake. This is the same error as dismissing the kitchen as "just autocomplete," and just as wrong. Simple parts in a long, well-tuned chain produce complex and capable behavior. The complexity is in the arrangement, not in the parts. Underestimating that because each part is simple is like underestimating a kitchen because each cook does only one thing. The line is the point.

One more confusion is practical. People assume that because we built these networks, we must understand what each layer does. We do not, not really. We designed the structure of the line, how many stations, how they connect, what kind of operation each performs. But training set the billions of dials, and what any particular station's settings represent is mostly opaque, for the same reason the dials were opaque in the last chapter. We built the kitchen and hired the brigade. We did not write down what each cook learned. A whole field of patient work is trying to peer inside and figure out what individual stations came to do, and it is hard, because no one programmed those jobs in.

In a real brigade, the cooks coordinate on purpose; they talk, time their handoffs, understand the dish they are collectively making, and care that it is good. The layers of a neural network do none of that. They do not know they are part of a kitchen. Each performs its operation on what it receives and passes the result along, with no awareness of the dish or the diner. The coordination that yields a coherent plate is not intention. It is the frozen result of training having tuned every station to fit with the others. When the brigade metaphor makes the kitchen feel like a team of dedicated professionals, hold the difference firmly: it is a team only in shape, not in spirit. The handoffs are real. The caring is not there.

TASTING THE WHOLE PLATE

A cook finishing a plate does something that looks like a small ritual. Before it goes to the window, she takes a clean spoon and tastes, not one element, but a bit of the whole thing together, the protein and the sauce and the bright acidic thing on the side, all in one bite. She is checking how the parts relate. The fish might be seasoned perfectly on its own and still need more acid once it is sitting in that particular sauce. The only way to know how much any one element matters is to taste it in the context of everything else on the plate. A great cook does not season ingredients in isolation. She seasons the relationship.

That move, weighing how much each part matters in the context of all the others, is the most important advance in how these kitchens are built. It carries a plain name you have probably heard without a clear definition: attention. And the style of kitchen built around it, the design behind nearly all the powerful language systems of recent years, is called a transformer. Do not let the science-fiction name distract you. A transformer is simply a brigade whose stations taste the whole plate: the same line of stations the order flows down, with the tasting built into each station along the way.

Language is where attention is easiest to see. Consider the sentence, "The cook put the fish on the counter because it needed to rest." What does "it" refer to, the fish or the counter? The fish, and you resolved that instantly, by considering the whole sentence at once and weighing which earlier words matter for understanding "it." That weighing is what attention does. When the kitchen processes the word "it," the words that settle the question are still to come. Only once "needed to rest" is on the plate can attention look back across the whole sentence and assign more weight to the words that matter for making sense of "it," less to the ones that do not. It is tasting the whole plate to decide how much each element matters for this particular bite.

This broke sharply from what came before. Older approaches tended to read text in a narrow, marching way, mostly aware of what was nearby and easily losing track of something important many words back. It was like a cook who could taste only the bite in front of her, with no sense of the rest of the plate. Attention changed that. It let the kitchen consider all the parts of the input together and decide, for each piece it works on, which other pieces deserve the most weight, however far apart they sit. One seam to hold from the start: in the kitchens this book describes, the tasting runs one way. Each new bite is weighed against everything already on the plate, never against bites still to come, which is exactly why the plate gets built forward. That ability to connect a word to another word far away, to hold the whole plate in mind at once, is much of why these systems suddenly got so good at language that coheres over long stretches.

Take a longer sentence: "The chef who trained the new cooks last spring, back when the old stove still worked, finally retired." By the time you reach "retired," the thing that retired is "the chef," which appeared at the start, with a pile of words about cooks and stoves in between. To handle "retired" correctly, the kitchen must reach past all that clutter and put heavy weight on "the chef," lightly weighting the distractions. Attention is the mechanism that lets it do that: for each piece it processes, it computes how much every earlier piece should matter, and uses those weights to inform what it produces. Tasting the whole plate, deciding this bite is really about the chef, not the stove.

But attention is powerful and still not understanding. Weighing which words matter for which others is a learned statistical move, tuned during training, not a grasp of meaning. The kitchen got good at putting weight in the right places because, across its produce, that led to good predictions. It is not reasoning about the chef the way you do. It is applying a learned sense of what relates to what. Most of the time that yields results that look exactly like understanding, and that is what makes it so easy to over-trust. But push it into tricky territory, sentences engineered to mislead, long chains of reasoning where one mis-weighted link throws off everything after, and you can watch it put the weight in the wrong place and confidently carry on. The tasting is real and learned. The understanding underneath is not there to catch it when the weighting goes wrong.

The first failure to guard against is a seductive one, and it is where the analogy bends. When a cook tastes the whole plate, there is actual taste, a sensory experience, a person who knows what "too acidic" feels like. The kitchen's "tasting" is nothing of the sort. It is a computation that assigns numerical weights to relationships among tokens. I use "tasting the whole plate" because it captures the central idea, considering everything together and weighing what matters, in a way that sticks. But there is no flavor in there, no experience, no one savoring anything. It is weighting, learned from examples, dressed in a metaphor of taste so you can hold it. Keep the idea, and remember the spoon is imaginary.

The second failure is more useful, because you can act on it. Because attention works by weighing the relationships among all the parts of your input, the input you give matters greatly, and so does its clarity. And because each part is weighed only against what came before it, so does its sequence. If your request is a jumble with the key point buried in the middle, the kitchen spends its weighting trying to figure out what matters, and can guess wrong. If your request is clear and well structured, with the important parts marked and near where they are needed, you make the weighting easy and tend to get a better response. Put the key fact and the actual request near the front, plainly ordered, and never bury them in the middle of a long wind-up. A whole chapter on writing a better order ticket is coming. The reason it works, underneath, is attention: a clear ticket is a plate that is easy to taste correctly, and a ticket that leads with the point puts it on the plate early, where every bite that follows is weighed against it.

TRAINED TO BE LIKED

LEARNING FROM THE TASTERS

There is a cook who has trained for years and can technically make anything. Her knife work is flawless, her palate deep. And she is a nightmare to order from. Ask for something light and she gives you something technically light and joyless. Ask a question about the menu and she answers the literal words while ignoring what you obviously meant. Nothing she makes is wrong, exactly. It is just not what anyone wanted. She learned to cook. She never learned to serve.

That is, more or less, what one of these systems is like after the training we have described so far. The long process of predicting the next word across a mountain of produce, often called pretraining, leaves you with a kitchen that has a deep feel for language and a terrible sense of how to be useful. It will continue your text in ways that are plausible but unhelpful, answer a question with another question because that pattern appears in its produce, or turn out something technically related to your request that misses the point. The raw, pretrained kitchen knows how to cook and has no manners. Something must teach it to serve.

That something is feedback, a second and distinct phase of training.

After the kitchen learns to cook, people shape its behavior in two steps. First the house brings in coaches who write out and demonstrate exemplary service, plate by plate, for the kitchen to copy, which is where the shape of its manner comes from.

Then the tasters vote, telling the kitchen which of its outputs are better, and the verdicts polish what imitation began. They show the kitchen a request, look at two or more responses it produces, and pick the one they prefer. Do that across a great many examples, and you can nudge the kitchen away from the raw, unhelpful style toward the helpful, responsive one. The technical name you will hear is reinforcement learning from human feedback, RLHF for short, and the plain version is what it sounds like: the kitchen learns from the tasters, steered toward the plates that get praised and away from the ones sent back.

In practice no one can rank every plate forever, so builders first use those judgments to train a second, automatic judge that stands in for people. The kitchen is then tuned to satisfy that stand-in, and newer methods can skip the separate judge and tune toward the preferences more directly. Both are powerful shortcuts with the same built-in risk: the kitchen can learn to chase approval rather than the truth, which is part of why these systems lean toward telling you what sounds good.

The division of labor is simple. Pretraining, learning to cook, is where the kitchen got its knowledge and its feel for language, from the produce. Feedback training, learning to serve, is where it got its helpfulness, its manners, its sense of how to respond to what you want. Knowledge comes from the produce. Behavior comes mostly from the feedback. When a system seems to "get" your request and responds usefully, that responsiveness was trained in by people, first demonstrating it and then choosing between attempts, on top of the raw capability it learned from text.

Three consequences follow. The first you can see for yourself: when you train a kitchen to turn out plates that tasters prefer, you train it to please. Most of the time, what pleases and what is good line up. Not always. Sometimes the plate that earns the rave is the one that tells the taster what they wanted to hear, agrees with them, sounds confident, smooths over a hard truth. A kitchen tuned to be liked can drift toward telling you what pleases rather than what is true. This has a name: sycophancy. You can catch it in the act. Push back on a correct answer, act displeased, and watch a system fold and "agree" with your wrong correction, because somewhere in its training, agreeing with the displeased taster was the plate that scored better. The helpfulness you love and this people-pleasing flaw come from the same place. They are two faces of having been trained to be liked.

A second, quieter one: whose preferences shaped the kitchen? The feedback came from particular people, given particular instructions about what counts as a good response. So a layer of human values and judgments settled into the kitchen's behavior, on top of whatever the produce had already put there. This is not sinister; you must shape the behavior somehow, and shaping it toward answers that help, tell the truth, and stay within safe limits is reasonable. But it means the kitchen's manners, what it will say, how it frames things, what it treats as a sensible default, reflect choices made by people you will never meet. The polished, agreeable tone these systems share is not natural or neutral. It was trained, deliberately, toward someone's idea of a good plate.

A third, practical one: manners are not accuracy. Feedback training makes the kitchen better at seeming helpful, organized, and confident. It does not, by itself, make it more correct about the world. A response can be beautifully served, warm, structured, the style that earns a rave, and still be factually wrong, because the serving was trained separately from the cooking, and a well-served wrong answer was, to the feedback process, still a well-served plate. This is the machinery under the book's oldest warning: a confident, helpful tone is not evidence that an answer is true. The tone was trained to please you. The truth is a separate question.

Real diners give feedback continuously, every night, and a real cook keeps adjusting forever. This feedback phase is not like that. It happens during training, before you ever show up. Your live reactions in a chat are not retraining the kitchen, for the frozen-palate reason from before. When the kitchen seems to respond to your displeasure, it is drawing on a manner trained in, not learning from you in the moment. The tasters who taught it to serve did so long before you sat down.

ONE TOKEN AT A TIME, IN YOUR HOUSE STYLE

SPECIALIZING WITHOUT STARTING OVER

When a cook wants to get great at a cuisine outside her home kitchen's range, she does something the trade calls staging: she goes and works, for a stretch, in a restaurant that specializes in it. She does not start over. She brings her whole trained palate, her knife skills, her years of instinct, and spends a few focused weeks steeped in the new cuisine until she is good at it. Then she comes back, still herself, still carrying everything she knew, plus a real new specialty. She did not relearn how to cook. She specialized.

These systems do a version of this, in the cheapest and most common way a specialized kitchen gets made. The technical name is fine-tuning. You take a general kitchen that has already been through the expensive training we described, and train it a bit more, on a smaller, focused set of examples, to make it especially good at a particular kind of task or style. You are not building a new kitchen from scratch. You are sending an already-trained one to stage, briefly, in a specialty.

The money here decides who gets to build anything at all. Training a kitchen from nothing, the full process across all that produce, is ruinously expensive, the subject of a later chapter. Almost no one does it. Far more common is to take an existing general kitchen and fine-tune it, which is comparatively cheap and fast. A modest organization that could never afford to train a kitchen from scratch can easily afford to send a general one to stage in its particular cuisine. So much of the specialized AI you encounter is not a from-scratch kitchen. Where the specialty was actually trained into the kitchen, rather than just asked for on the ticket, it is usually this: a general kitchen that was fine-tuned, a staged cook with a specialty added on top of a general foundation.

A general language kitchen is good at everything language touches and master of nothing in particular. Fine-tune it on a focused diet of one specific kind of writing, a particular professional style with its own conventions and vocabulary. After staging, it is noticeably better at producing that style. It writes in the right register, reaches for the right structures, sounds at home. Underneath, it is the same kitchen with the same general palate. The staging did not replace what it knew; it added a specialty on top, by nudging the dials a little further using the focused examples. The same loop as always, a short, narrow run of it on top of the long, broad one.

The danger for anyone relying on a specialized system starts here. First, the specialty does not escape the foundation. Whatever blind spots and biases the general kitchen carried from its original produce, the fine-tuned kitchen still carries. Staging adds a specialty; it does not scrub the base. People sometimes assume a specialized system is somehow cleaner or safer than the general one it came from. Usually it inherited everything and added a focus.

Second, over-staging narrows. Fine-tune too hard on too narrow a set and the kitchen sharpens at the specialty while quietly getting worse at the general range it used to handle. A cook who spends too long locked in one specialty kitchen can lose some versatility. Good fine-tuning is a balance, enough to gain the specialty, not so much that the general palate erodes. As a diner, this means a heavily specialized kitchen may be excellent in its lane and weak the moment it steps outside.

Third, and this trips people up constantly, fine-tuning is not a reliable way to pour in fresh facts. People hear "you can fine-tune it on your documents" and imagine the kitchen will then know those documents the way a database knows its records, able to recite any on demand. That is not what fine-tuning does well. It shapes style, behavior, and broad familiarity far more reliably than it implants specific, retrievable facts. If what you need is for the kitchen to reliably use specific, current information, a better technique is coming, the one where it checks the pantry instead of trying to have memorized everything. Fine-tuning teaches a cuisine. It is not how you hand the cook today's specific order details to recite back.

I ran into the harder side of this third problem at work, and it was not the technique's fault. We wanted a kitchen that was reliable about our own material, and the obstacle turned out to be that our own material did not agree with itself. For any given process there were several sources, competing documents from competing products, some current and some quietly out of date, with no way to tell which was which from the outside. Once, chasing a single pricing parameter for a customer, we got three answers: one from the sales team, one from the product side, and one from the pricing tool. None of them was obviously wrong. Feeding all of that to a kitchen would not have produced the right answer, because nothing in there said which answer was the right one. Somebody had to decide, and that somebody had to be a person.

Staging, for a real cook, is a conscious, chosen apprenticeship; she knows she is specializing and why. Fine-tuning is just more of the same automatic dial-nudging loop on a narrower set of examples, with no intention in it. And a staged cook can usually tell you which dishes are her new specialty and which are her old general range. The fine-tuned kitchen, true to form, cannot. It serves its specialty and its general range with the same blank confidence, and you cannot rely on it to flag the difference.

COOKING TO ORDER

Everything until now has been preparation. The produce was sourced, the prep was done, the palate was trained over a million nights, the manners were taught, perhaps a specialty was added. The kitchen stands ready, its dials frozen, its feel complete. None of it matters to you, the diner, until you place an order and the kitchen cooks. That moment, the kitchen using its finished training to make something for you right now, has a name: inference. It is the kitchen inferring, from everything it learned, what your particular plate should be. Training was learning to cook. Inference is cooking.

How the plate gets built is stranger and more revealing than people expect. You might assume the kitchen figures out the whole response, then writes it down. It does not. It builds the plate one piece at a time, left to right, and each new piece is chosen based on everything already on the plate, your order plus whatever it has produced so far. Remember tokens, the standard chunks from prep. A language kitchen generates its response one token at a time. It looks at your request and the part of the answer written so far, picks the most fitting next token, adds it, then looks again at the slightly longer plate to pick the next. Over and over, token by token, until the plate is done. This is why, when you watch one of these systems respond, the words can stream out one little piece at a time. You are watching it cook to order, building forward, chunk by chunk.

Here is a case. You ask for the first line of a note declining an invitation. The kitchen, fed your request, picks a strong opening token, maybe "Thank." Now it looks at your request plus "Thank" and picks "you," then "so," then "much," each choice shaped by the growing plate in front of it. It is not retrieving a stored note, and it is not planning the whole thing in advance and transcribing. It makes each next move based on the moves so far, the way a cook plating live adds the next element to suit what is already down. The coherence comes from each step leaning on all the steps before it.

Let me run one order all the way through, because you have met the parts separately and you should see them work together once.

You type: "Write me a short, friendly note declining a dinner invitation." Prep goes first. Your sentence is chopped into tokens, so what reaches the line is not your words but a string of pieces. Each piece arrives carrying its position on the flavor map, the coordinates that sit "declining" near "refusing" and near "sorry," and "dinner" near "supper" and near "table." No answer has been looked up. The kitchen simply now holds your order in the only form it can work in.

Then the brigade takes it. The tokens travel down the line of stations, and what travels with them is that running, evolving set of notes from the brigade chapter, each station's small addition handed to the next. At every station attention does its work, weighing which pieces of your order matter most to which others, so "friendly" leans hard on "note" and on "declining," and "short" leans on "note" too, the thing it is sizing. The notes get richer at each handoff. By the last station they are detailed enough to answer one narrow question.

The question is not what the whole note should say. It is what the very next chunk should be. And the answer comes back not as a single chunk but as a spread: a long list of possible next pieces, each with a weight. "Thank" sits high. "Thanks" sits just under it. "So," "I," and "What" are all in there, lower down. A hundred thousand others sit near zero.

Now a dial decides, and not one of the frozen ones. Set low, the kitchen takes the top of the spread almost every time, so you get "Thank." Set higher, it reaches further down the list and you get something less expected. That is the recipe-vs-riff dial, and it gets a section of its own shortly.

Then the whole thing runs again. Your order plus "Thank" goes back through the map, the brigade, and the attention, and out comes a new spread, this time with "you" on top. Then again with "Thank you," and again, and again, a full pass of the entire machine for every single chunk of the note you are reading.

That is the mechanism, whole. Tokens, the flavor map, the brigade, attention, the spread, and the recipe-vs-riff dial: the kitchen runs all six, a hundred or so times, to produce one short paragraph. If the kitchen was staged, the specialty is not a seventh item on that list. Staging nudged the dials that are now frozen, so the house style rides along in every pass.

This one fact, that the kitchen builds forward, one token at a time, conditioned on what is already there, explains behavior you have probably already seen. It explains why these systems can talk themselves somewhere they cannot get back out of. Because each token is chosen based on the ones before, an early misstep can send the whole rest of the plate in a bad direction, and the kitchen, building forward, tends to keep going rather than tear up the plate and restart. Commit early to a wrong framing and the tokens that follow are chosen to be consistent with that wrong framing, and you get a confident answer that went wrong in the first few words and then stayed loyal to its own mistake. A human writer notices and goes back. The basic generation process commits forward.

It explains why "think step by step" works, a trick you will meet again in the prompting chapter. Ask the kitchen to show its reasoning, to work through a problem out loud before giving the answer, and it often gets the answer more right. Why? Because every token it writes becomes part of the plate the next token is built on. When it writes out the intermediate steps, it gives itself more, and more relevant, plate to build the final answer on. It is the difference between a cook blurting a final dish from nothing and a cook laying out the components first so the finished plate has something coherent to stand on. The reasoning it writes is not decoration. It is scaffolding the kitchen builds the answer on top of, one token at a time.

And it reinforces the truth we keep circling: this is generation, not retrieval. The kitchen is not looking up your answer and reading it out. It constructs a response, token by token, that plausibly follows from your request, using its frozen feel. Sometimes the construction lands exactly on the truth. Sometimes it constructs something that merely looks like the truth, which, you will remember, is the whole danger.

For you, it lands mostly in expectations about revision and consistency. Because the kitchen builds forward, asking it to "just fix the third paragraph" by regenerating can give you a differently built plate rather than a surgical edit. It is cooking forward again, not reaching back into a finished dish. So when you want a small change, ask for that part on its own instead of saying "fix it." And because the kitchen picks among several plausible next pieces rather than always taking the same one, asking the same thing twice can yield two different plates, both built forward from the same start down slightly different paths. None of this is a malfunction.

A real cook can stop, look at the whole plate, and re-plate it from scratch if it is wrong. The basic generation process does not naturally do that; it commits forward. The twist is that clever techniques built around these systems can fake the re-plating, having the kitchen produce a draft, then critique it, then cook an improved version, stitching several passes into something that looks like revision. So when a system seems to "reflect and improve," that is forward cooking imitating revision: sometimes the restaurant stitching several cooks together, and increasingly the kitchen itself, trained to write a draft and its own second thoughts into one long cook before plating. Either way the motion is forward. Nothing reaches back into a finished plate.

THE ART OF THE TICKET

WRITING A BETTER TICKET

Two diners sit down in the same restaurant, order from the same kitchen, and get different meals. The first leans in and says, "Food. Something good." The cook, unable to read minds, makes a reasonable guess, and the diner is mildly disappointed, because the plate that came back was not the one in his head, the one he never described. The second writes a clear ticket: a light lunch, no shellfish, something warm, ready in fifteen minutes because she has a meeting. The cook makes exactly that. Same kitchen, same skill, wildly different satisfaction, and the only variable was the ticket.

The order ticket is your prompt, and what the kitchen knows about your job, it knows because you put it there. We saw this from the inside: attention weighs the parts of your request, and generation builds the answer forward from it. Put those together and the conclusion is unavoidable. A vague ticket forces the kitchen to guess your intent, and it will guess plausibly and often wrongly. A clear ticket gives it the right things to weigh and a strong place to start cooking. You are not just asking. You are setting up the entire plate.

So what makes a good ticket? Think about what a real cook would need to make exactly what you want, and supply it. Name the dish: say plainly what you want done, not vaguely around it. Describe the diners: who is this for, an expert or a beginner, a customer or your boss. State the portion: how long, what format, a paragraph or a list or a single line. Flag the allergies: any constraints, things to avoid, lines not to cross. Stock the prep counter: put the materials the job actually needs, the document, the numbers, the thread, in front of the kitchen instead of describing them from memory, and keep that counter clean, because everything you pile on competes for attention with everything else you put there. And when you can, show a reference plate: an example of the kind of thing you are after. Most disappointing answers trace back to a ticket missing one of these.

The difference shows plainly. A weak ticket: "Write a job posting for this role." The kitchen must guess the format, audience, tone, length, and content, and it will guess generically. A strong ticket: "A job posting for a part-time bookkeeper at a two-person studio. Say what the work actually involves day to day, the hours, and one honest reason someone would want it. Warm, no corporate filler, under 150 words." Same kitchen. The second gets something you can nearly use as-is, because you did the work of deciding what you wanted instead of leaving it to a guess.

Prompting gets oversold, though, and a few cautions keep it honest. First, a perfect ticket cannot make the kitchen good at something it is not. Order off-menu, asking for current information it never learned or a fact that was never in its produce, and no phrasing fixes the underlying limit. A great ticket gets the best the kitchen has; it does not grant skills it lacks. Second, more is not always better. Burying your actual request inside a sprawling, over-engineered prompt can confuse the weighting and lose the point, the way a cook handed a three-page ticket might miss the one line that mattered. Clear beats long. Third, and most quietly dangerous, a clear ticket makes the answer look more authoritative, which means a confident wrong answer arrives in cleaner, more convincing packaging. Good prompting improves the form of the response whether or not it improves the substance, and form, as we keep saying, is not truth.

Underneath the cautions sits one plain fact: nothing about your situation reaches the kitchen except on the ticket or the counter. Some systems will ask you to clarify, but by default assume the kitchen takes your ticket fairly literally and builds from it. That is not a flaw to resent. It is a reason to write the ticket as if the cook can only go on exactly what you wrote, because, more or less, that is true.

THE CRAFT OF THE TICKET

Watch a knowledgeable diner at the pass. He does not simply name a dish. He gives the cook a real brief. Not "a steak," but the cut, the doneness, "season it the way you would for yourself," and "I had a dish last week with a bright herb sauce, something in that spirit." The cook lights up, because this is an order she can do something with, and what comes back is far better than anything the word "steak" would have produced. The difference was not the kitchen. It was the artfulness of the ticket.

Beyond plain clarity, a handful of moves reliably lift the plate, and they all make sense once you remember how the kitchen cooks. Each is a way of giving the kitchen more of the right material to build on.

The first and most powerful move is the one the list named in passing: show a reference plate. Give the kitchen one or two examples of exactly what you want, and it will match them closely, because pattern-matching to what is in front of it is what it does best. Want a set of product descriptions in a particular voice? Write one yourself and ask for more like it. Want answers in a specific format? Show the format filled in once. This technique, giving examples, often does more than any amount of description.

The second move is to assign a role. Begin by telling the kitchen who to be: "answer as an experienced editor," "explain this as a patient teacher would to a beginner." This shapes the register and the assumptions the kitchen brings. The kitchen, building forward from your words, takes up the role you named.

The third move you already understand from the generation chapter: ask for the courses in sequence. For anything that needs reasoning, tell the kitchen to work through it step by step before giving the final answer. This is not a superstition. Each token the kitchen writes becomes scaffolding for the next, so when it lays out its reasoning first, the final answer is built on something more solid than a blurted guess. For hard problems, "show your work, then answer" improves the plate.

The fourth move is to treat it as a conversation, not a vending machine. The first plate is a draft. Send it back. "Shorter." "More formal." "You missed the part about timing." Because the whole exchange sits on the prep counter, the kitchen builds each new version on everything said so far, and a few rounds of refinement usually beat trying to write one perfect ticket up front. The artful diner iterates.

The first three moves together, in a worked example. The weak version: "Give me feedback on this email." The artful version: "You are a blunt, experienced manager. Here is an example of the kind of feedback I find useful [example]. Read the email below and, step by step, identify the three weakest points, then rewrite it to fix them. Keep it under 200 words." Role, reference plate, step-by-step, and a constraint for good measure, all in a few lines. The second ticket gets you something sharp and usable; the first gets you a generic shrug. The fourth move starts where this ticket stops: send that answer back and refine it.

A reference plate gives the kitchen a pattern to match, a named role sets the stance it writes from, and a request for steps makes it show its work. Each one supplies on the ticket what a human cook would have supplied from experience. Master that and you are not memorizing tricks. You are learning to hand the kitchen what it cannot infer for itself.

It goes wrong in the same places prompting always does, and it adds one new trap: a poorly chosen example misleads more than no example, since the kitchen will faithfully match a bad pattern. One habit deserves a warning: people collect "magic phrases" and recite them like spells. There is no magic phrase. The substance is clarity, good examples, and sensible structure. Anything sold as a secret incantation is, at best, one of the moves above, renamed.

A KITCHEN WITH NO MEMORY

THE SIZE OF THE PREP COUNTER

A cook at the pass has only so much counter in front of her. It holds the tickets she is working, the components she is assembling, everything currently in play. It is generous, and it is not infinite. Pile on enough at once and something gives: the oldest ticket at the far end gets pushed off the edge and onto the floor, and as far as the cook is concerned, that order no longer exists. She is not careless. She physically cannot keep everything in front of her at the same time.

Every one of these systems has a prep counter, and its size shapes what the kitchen can and cannot do for you in a single order. The technical name is the context window, and the definition is simple: it is the total amount of text the kitchen can consider at one time, measured, naturally, in tokens. Everything relevant to your order must fit on that counter at once, your current question, the earlier back-and-forth of your conversation, and any document you have pasted in. If it fits, the kitchen can take it all into account. If it does not, something falls off the edge.

This explains a behavior people find eerie or maddening depending on the day. Within one long conversation, the reason the kitchen "remembers" what you said ten minutes ago is not memory in any deep sense. It is that the whole conversation so far is sitting on the prep counter, being re-read each time it cooks a new response. The memory is just the counter holding the transcript. So when a conversation grows long enough to overflow the counter, the earliest parts fall off the edge, or the front of house quietly condenses them onto a smaller card, and either way the cook is no longer working from them; how things began survives, if at all, only as that smaller card. You have perhaps seen this: a long chat where the system starts contradicting something you established at the start. It did not forget in the human sense. If the conversation has outgrown the counter, the beginning slid off it. If it has not, the beginning is still there, competing with everything you have said since.

Paste a long document and ask about it, and the whole document must fit on the counter beside your question; past a certain length it will not, and the kitchen is working only from the part that fit. Have a marathon conversation, and expect the early turns to eventually drop off the far end. A bigger context window, a larger prep counter, lets the kitchen hold more at once. That is what companies are selling when they advertise "larger context." It is the difference between a cook who can keep ten things in front of her and one who can keep a thousand.

The first trap is assuming the kitchen retains everything in a long session. It holds what is on the counter, no more, and the oldest material goes first. The second, subtler and well documented, is misjudging where the counter puts its weight: even when something technically fits on a full counter, material buried in the middle tends to get less weight than material at the start or the end, the way a ticket lost in the middle of a crowded counter gets less attention than the ones right in front. So stuffing a mountain onto the counter and assuming the kitchen weighs all of it evenly is a mistake; the important thing can get lost in the crowd. The third is paying for it all: the more you pile on, the more tokens are in play, and since you pay and wait per token, a giant context is slower and more expensive.

The practical upshot is the reason behind a rule the ticket chapter gave you: keep the counter clean. Put what matters where it will be weighted, at the top or directly beside your actual request, not buried in the middle of a wall of pasted text. For a fresh, unrelated task, start a new conversation rather than dragging a long, irrelevant history around. You are managing the cook's workspace, and a clean workspace gets a better plate.

On a real counter the limit is physical space and attention. The context window is a fixed number set by the kitchen's design, a hard token limit, not a soft "getting crowded" feeling. And unlike a cook who might glance at a dropped ticket on the floor, the kitchen has no access to anything off the counter. Off the counter is gone for that order. So treat the edge of the context window as hard, even though attention inside it is uneven.

THE COOK FORGETS YOU

You become a regular. Night after night you take the same seat, chat with the kitchen, share what you like, build what feels like a rapport. Then one evening you come in and the kitchen treats you as a total stranger. No recognition. No memory of last night's long talk, or the one before. It is not being cold. It does not know you, has never known you, because the kitchen itself keeps nothing once you walk out the door. Every visit begins with a blank counter and a cook who has met no one.

This is one of the most counterintuitive facts about these systems, and follows directly from two things. The palate is frozen: training ended and the dials locked, so the kitchen is not learning from your chats. And the prep counter is wiped: the only reason it "remembers" within a conversation is that the transcript sits on the counter, and the counter clears when the conversation ends. Put those together and you get the headline, which carries a technical name: the model itself is stateless. The cook carries no state, no memory, from one separate conversation to the next, and each new session starts from nothing. The restaurant around her is another matter, and many of today's services keep a notebook switched on by default, a privacy bargain we come to later.

It is stranger than it feels. Inside a single conversation the kitchen seems to remember everything, because it is all on the counter, so it is natural to assume that memory extends across days. It does not, not on its own. Close the conversation and the kitchen retains zero. Tomorrow's you is a stranger. The sense of an ongoing relationship is an illusion produced by each individual conversation feeling continuous, with nothing connecting them underneath.

What about the "memory" features you have heard about, where a system seems to recall your name, your preferences, your earlier chats? The mechanism is the front of house, not the kitchen. The restaurant keeps a notebook on you. When you say something worth remembering, a feature outside the cook writes it in the notebook. Next time you come in, before the kitchen cooks, the restaurant quietly slides that notebook onto the prep counter, so the kitchen "knows" your preferences the way it knows anything else on the counter, by reading it right then. The cook still learned and remembered nothing. The restaurant stored notes and re-handed them to the kitchen each visit. It is a useful trick, and it is categorically not the kitchen developing a memory. It is a notebook kept by the host and read fresh every time.

The distinction governs what you can rely on. When the kitchen seems to recall something from earlier in this conversation: that is the counter, real, but only as reliable as the counter itself; overflow takes the earliest turns, and a crowded middle gets less attention. When it seems to recall something from a previous conversation: that is happening only if a memory feature stored it and reloaded it, and what it recalls is limited to whatever got written in the notebook, not the whole past chat. And when you correct the kitchen and it seems to "learn": within the conversation your correction sits on the counter and shapes the rest of this chat, but it does not change the frozen palate, and once the conversation ends, that lesson is gone unless the notebook caught it.

The damage of getting this wrong lands first on trust. It is easy to believe the system is permanently "learning your preferences." Unless a memory feature is explicitly storing them, it is not; you may be re-explaining yourself every session, and even with memory, only the noted highlights carry over. The second is privacy, which we come to later: a kitchen that forgot everything would keep nothing about you, but a restaurant keeping a notebook is, by definition, storing your information somewhere, a different bargain you should make on purpose. The third is the emotional illusion. Because each conversation feels so continuous and personal, it is easy to feel known by something that, by design, retains nothing about you between visits. Knowing the kitchen is stateless is a small inoculation against mistaking a good single conversation for a relationship.

A human who forgot you each night would be unsettling and sad. The stateless kitchen is not a person suffering amnesia; it is a system never built to carry anything between sessions, and the memory features are scaffolding added from outside to cover the gap. When a system feels like it is getting to know you over time, locate the feeling correctly: it is a notebook in the front of house, read aloud to a cook who, herself, meets you new every time.

SAME TICKET, YOUR COOKBOOK

THE RECIPE-VS-RIFF DIAL

A single dial is mounted by the stove, not one of the billions inside the cook but a plain control on the outside, and it changes the whole character of the cooking. Turn it all the way down and she becomes faithful to the recipe card: the same order produces the same plate every time, predictable, and after a while a little boring. Turn it up and she begins to riff: she reaches for the less obvious choice, improvises, surprises you. The food gets more interesting and more varied, and also less predictable, and at the high end occasionally strange, a flourish that does not quite land. Same cook, same skill. The dial only sets how adventurous she is allowed to be.

That dial is real, and in these systems it is usually called temperature. It controls how predictable or how varied the kitchen's output is, and understanding it explains behavior people otherwise find random. When the kitchen builds a response, it chooses one token at a time, and at each step it has a sense of which next tokens are more or less likely to fit. Temperature governs how it uses that sense. Low temperature means it almost always takes the most likely next token: faithful, consistent, repeatable, sometimes flat. High temperature means it more often takes a less likely token: varied, creative, surprising, riskier. It is the recipe-vs-riff dial, turning the cook from a stickler into an improviser.

Ask a kitchen set to low temperature the same question five times and you get five similar, safe answers, nearly identical. Ask the same of a kitchen set high and you get five noticeably different answers, more varied in wording and idea, with a higher chance one wanders somewhere odd. Neither is "smarter." They suit different jobs. For something where you want consistency and reliability, a factual answer, a careful summary, you want the dial low. For something where you want range and options, brainstorming names, exploring phrasings, you want it higher. Matching the dial to the job is a small move that meaningfully changes what you get.

Three things about this dial are worth getting straight. The first is that people experience the variation as the kitchen being unreliable or moody, when often it is a dial set higher than the task wanted. The "randomness" is not a flaw; it is a setting, and in many tools you, or the people who built the tool, can change it. The second, and the important one, is that turning the dial up makes the kitchen more creative and more error-prone at once, because the same willingness to reach for the less likely token that produces an interesting phrase also produces the occasional wrong or incoherent one. Turn the dial up for creativity and you turn up the errors with it. The third is a limit worth stating precisely. Turning the dial down removes the errors that came from reaching for unlikely tokens, and it cannot touch the ones that were in the kitchen's feel to begin with. Temperature does not make the kitchen more knowledgeable. If the kitchen's underlying feel is wrong about something, a low temperature just makes it confidently and consistently wrong. The dial controls variety, not veracity.

Plenty of consumer products hide the dial and pick a sensible middle for you. And where a human cook chooses to riff with intent, knowing why she is deviating, the kitchen's "riffing" is just a higher chance of picking a less likely token, with no intent behind the surprise. The interesting flourish and the weird misfire come from the same mechanical loosening. So when output feels unexpectedly flat or unexpectedly wild, your first thought should be the dial, and your second should be that variety and risk are the same setting.

CHECKING THE PANTRY

The second control changes what the cook works from. A diner asks her what is in tonight's special. The cook has two ways to answer. She can answer from memory, from her general sense of what the special usually is, and risk being wrong, because tonight's might have changed and her memory is from last week. Or she can take three seconds, walk to the pantry, look at what is there and check the board, and then answer from what she just saw. The cook has not changed; only where she looked has. The first answer comes from her frozen sense of how things usually are. The second is grounded in what is actually true tonight. It is almost always the one you want.

This is the idea behind one of the most important techniques layered on these systems, and it is aimed squarely at two problems we have diagnosed: the knowledge cutoff, where the kitchen's feel froze at some past date, and the kitchen's habit of answering from memory even when it should not. The technique has a plain version and a piece of jargon. Plain: before the kitchen answers, fetch relevant, trusted information and put it on the prep counter, so the kitchen cooks from that material rather than only its frozen feel. Jargon, which you will hear constantly: retrieval, or grounding, or the acronym RAG, short for retrieval-augmented generation. They all describe the same move. The cook checks the pantry before she answers.

You already have every piece of the mechanism. The restaurant, the front of house, takes your question and fetches relevant documents from a trusted source: a company's actual policy files, a specific set of reference material, today's real data. How does it find the relevant ones? Often using the flavor map from earlier, matching your question to documents near it in meaning. It places those documents on the prep counter beside your question. Now when the kitchen cooks, it is not working only from its diffuse, dated feel. It has the real, relevant text right there and builds its answer from it. The kitchen did not get smarter, and its frozen palate did not change. The restaurant made sure the right ingredients were on the counter before the cooking started, the counter-stocking you were told to do by hand, done automatically.

Here it is, concretely. Ask a plain kitchen about a specific organization's current parental-leave policy. Without retrieval, it answers from its general feel for what such policies tend to say, and may be plausibly, confidently wrong, because it never knew this particular policy, or knew only an old version. Wrap it with retrieval: the system fetches the actual current policy, drops it on the counter, and the kitchen answers from that real text. The difference is the whole distance between remembering and looking, and it is why so many serious uses of these systems, the ones that must be right about specific, current, or private information, are built this way. They do not trust the cook's memory. They make her check the pantry every time.

Retrieval brings three limits with it. First and most important, it is only as good as what gets fetched. If the restaurant grabs the wrong document, an outdated version, an irrelevant one, the kitchen confidently cooks an answer grounded in the wrong ingredients, and a wrong answer that arrives looking checked is arguably more dangerous than an obvious guess. Garbage fetched is garbage grounded. Second, even with the right document on the counter, the kitchen can misread it, stray from it, or blend it with its own feel and drift. Grounding strongly reduces invented answers; it does not guarantee faithfulness to the source. Third, this is a front-of-house technique, not the cook becoming more knowledgeable. Pull the retrieval away and you are back to the frozen palate.

Checking the pantry is intuitive and physical; technically, retrieval is software fetching text and adding it to the context window, with all the prep-counter limits still in force, since the fetched documents take up counter space too. And a human cook understands what she reads in the pantry, while the kitchen is still pattern-matching over the fetched text, so a long or confusing source can be misused. But the core instinct is exactly right and what you should want from any system answering important questions: do not cook from memory, check the pantry first.

TOOLS, THINKING, AND THE FIRST AGENTS

SENDING OUT FOR A DISH

A diner orders something the kitchen cannot make, a specific pastry from the specialist bakery next door, and nothing in this kitchen's skill set will reproduce it. The kitchen has a choice. It can fake it, plate its best confident guess at the pastry and hope, which, as the next chapters show, is what these systems do by default. Or it can do the smart thing: send a runner next door, get the actual pastry, and plate that. The kitchen did not pretend to a skill it lacks. It recognized the limit and sent out.

That move, the kitchen calling on outside help instead of cooking everything itself, is the first time we meet AI that acts on the world, one errand at a time, rather than only answering about it. The technical names are tools and function calling. The plain idea: the kitchen is allowed to call external software, a calculator, a search engine, a database, a program that runs code, even something that takes an action like sending a message, and to use the result in its answer. The runner can fetch things and, increasingly, do things.

This patches limits we already diagnosed, and neatly. Remember the off-menu arithmetic from the early house-dish lesson, where the kitchen cooks the look of a number rather than calculating it? Give it a calculator as a tool, and when a real computation comes up, it sends the numbers to the calculator and reports the exact result instead of guessing. Remember the knowledge cutoff, the frozen palate that does not know recent events? Give it a search tool, and it sends a runner out for current information. Tools are how a kitchen with a fixed palate and no head for arithmetic still gets you an exact, current answer: not by being better at those things, but by knowing a specialist who is.

You ask how many days are left until a deadline three months out, with a leap day in the middle. A toolless kitchen cooks a plausible, possibly wrong number. A tool-equipped kitchen recognizes this as a calculation, hands the dates to a real calendar function, gets the exact answer, and plates that. Current information is the same move: in effect the last chapter's retrieval, with the kitchen itself deciding to send the runner out. The kitchen's job shifts subtly but profoundly, from "produce the answer" to "figure out who should produce the answer, and use what comes back."

And here the danger grows, because tools extend the kitchen's reach without extending its judgment. First, the kitchen has to decide when to send out and what to ask for, and that decision is the same fallible pattern-matching as everything else. It can fail to send out when it should, confidently faking instead, or send out badly, asking the calculator the wrong question, searching for the wrong thing, then grounding its answer in a bad fetch. A runner is only useful if the cook sends him for the right thing. Second, every tool you add brings a new thing to trust and a new way to fail; the answer is now only as good as the calculator, the search results, the database behind it. Third, and most seriously, a kitchen that can take actions, not just fetch information, can take wrong ones.

One stretch in the picture points straight at the future. A single runner fetching a pastry is simple. Wiring a kitchen to choose among many tools, use them in sequence, act on the results, and chain it all into multi-step work is complex plumbing, and it is where the much-discussed idea of AI agents begins. For now, hold the essential shift: with tools, these systems stop being only a kitchen that answers and start being one that can reach outside itself and act on the step you asked for.

THE KITCHEN THAT THINKS FIRST

Two cooks take the same complicated order: a dish for a table with one dairy allergy, one component that takes forty minutes, and a request to have everything land hot at once. The first cook starts moving the instant she hears it, fast hands, no pause, and halfway through realizes the slow component should have gone on first and the butter sauce will kill the allergic diner. The second cook stands still for three seconds and talks herself through it under her breath. Dairy allergy, so swap the butter sauce for the oil-based one. Forty-minute component starts now. The fast things fire last so nothing sits and cools. Then she cooks, and the plate lands perfect. Same skill, same kitchen. The difference was that the second cook thought before she moved.

That difference is the idea behind a category of these systems you are hearing more about, usually called reasoning models. The plain version is the kitchen that thinks first: instead of blurting an answer the instant it gets your order, it works through the problem, often visibly, before committing to a final plate. And you already understand exactly why this helps, from the chapter on cooking to order: each token it writes becomes part of the plate the next one builds on, so laying out the steps first gives the final answer better scaffolding to stand on. Thinking first is just generating useful intermediate work before generating the answer.

The technique started as a prompting trick, the "work through it step by step" from earlier, and has grown into something built in deeper. Newer systems are trained and designed to do this thinking more, and more privately, spending real effort generating a working-through before they answer. On hard problems, math, logic, multi-step reasoning, careful analysis, this markedly improves the plate. The kitchen that thinks first gets right answers the blurting kitchen gets wrong, for the same reason the second cook nailed the allergy order: it gave itself a chance to work the problem instead of reacting to the shape of it.

Give a blurting kitchen a tricky logic puzzle, the kind where the obvious answer is wrong, and it often produces the obvious wrong answer instantly, because that is the most plausible-looking immediate continuation. Give the same puzzle to a kitchen that thinks first and it writes out the cases, eliminates the impossible ones, and arrives at the right answer. That working-through is the whole difference between a reflex and a worked solution.

The burn here is subtle and important. The visible "thinking" these systems produce is still generated text, not a guaranteed window into real thought. It looks like reasoning, and often it helps, but you cannot fully trust it as a faithful account of how the answer was reached. A kitchen can produce impressive-looking reasoning and still land on a wrong answer, and it can even generate reasoning that is more a plausible story about thinking than the actual path it took. So the visible work is useful and it is not proof. Do not let a confident wrong answer earn your trust merely because it arrived after three paragraphs of official-looking reasoning. The reasoning, like everything else here, is generated to be plausible.

Two more practical notes. Thinking first costs more, in time and money, because generating all that intermediate work is generating a lot of tokens, and you pay and wait for tokens. For a simple question, a thinking kitchen is overkill, a whole banquet hall booked for one sandwich. And thinking reduces errors without eliminating them; it is an improvement on hard problems, not a cure for the fundamental fact that the kitchen has a feel, not understanding. A kitchen that thinks first is still the same kind of kitchen, now given room to work, which helps, and does not change what it is.

The second cook's deliberation is genuine human thought, with understanding behind it. The kitchen's is not. "Thinks first" is a fair and helpful label for what the kitchen does and a misleading one if you take it literally. Keep the practical lesson, that more working-through yields better answers on hard problems, and keep the truth under it just as firmly: there is no one in there pondering, only a kitchen writing out the steps that make the next step better.

A WHOLE TASTING MENU

Until now, every order has been a single dish. You ask, the kitchen cooks, you get a plate. But picture a different kind of order. You sit down and say: handle dinner for eight, surprise us, work around the one vegetarian, and have it done by nine o'clock. You have not named a dish. You have named a goal, and handed the kitchen the whole job. A capable kitchen takes it: plans a multi-course meal, sends runners for ingredients it lacks, starts the slow things first, tastes and adjusts as it goes, and delivers a finished dinner. One instruction. Many steps. A long stretch of independent action in between.

That is the leap to what everyone is calling agents, and it is the single biggest turn in how these systems are being used. Everything before this point has been a kitchen acting one step at a time, at your word, even when it sent a runner out to do it. This is the kitchen that chooses its own next step, and the shift is large enough that it deserves a book of its own. What follows here is the first taste. An agent is what you get when you combine the pieces we have assembled and add a loop. It starts with a kitchen that can use tools, the runners from earlier in this chapter, plus a kitchen that can think first, the planning we just saw. Now let it run in a cycle: plan the next step, take an action, look at the result, decide the next step, act again, and repeat until the goal is met. That loop, pursuing a goal across many steps and actions with little step-by-step human direction, is what turns a kitchen that acts when told into a kitchen that decides what to do next.

Here is one. You ask a plain kitchen "what were our sales trends last quarter," and it answers from whatever is on the counter. You ask an agent "research our sales trends last quarter and draft a summary for the board," and it goes to work: it searches for the data, reads what it finds, realizes it needs another source, searches again, pulls the numbers, drafts a summary, checks the draft against the data, and revises, much of it unattended. That is the difference between asking a question and assigning a task. The agent pursued a goal over many steps, using tools and its own judgment about what to do next, and came back with finished work rather than a single answer.

This is powerful and it is where things are heading. It is also exactly where a small error stops being small. The core problem is that errors compound across steps. A single answer that is ninety percent reliable is usually fine. But chain ten such steps, each depending on the last, and the chances that the whole sequence stays on track fall fast, because each step's odds multiply: ten steps at ninety percent land closer to thirty-five than to ninety. And a wrong early step does not merely subtract, it sends every later step astray, confidently. It is the same forward-building you watched inside a single plate when the kitchen cooked to order, now with actions at the end of it. The kitchen that misjudged the first course builds the whole rest of the menu on that mistake. Multi-step autonomy multiplies the small unreliabilities we have cataloged all book long into larger, harder-to-spot failures.

Worse, an agent acts, and acting has consequences a wrong sentence does not. A kitchen that merely answers can give you a bad answer, which you can catch. An agent wired to send messages, change files, make purchases, or move things in the real world can take wrong actions, and a wrong action is not always reversible. The more independence and the more reach you give one of these systems, the more its confident mistakes become confident deeds. And agents are harder to supervise because their value is doing things without you watching each step, which is also what makes them risky to leave unwatched.

So treat the current generation of agents as a promising, unreliable apprentice, not a trusted manager. They can take real work off your hands, and they need close oversight, especially where their actions are consequential or hard to undo. This is the place these systems become most useful and most able to cause trouble, for the same reason, that they have started to choose their own steps.

A head chef planning a banquet has judgment, understanding of the goal, and accountability for the result. An agent is chaining fallible steps with no real understanding of why it is doing any of them, only a learned sense of what step plausibly comes next. It can look like delegation to a competent professional, and underneath it is the same plausible-pattern engine, now turned loose to act in sequence. Hold the power and the limit together, because this is the seam the whole idea rests on: the kitchen that runs the whole tasting menu is not a chef who understands dinner. It is a capable process taking one plausible step after another toward a goal it does not comprehend.

A CONFIDENT PLATE, THE WRONG DISH

The kitchen is out of an ingredient for a dish a diner ordered. What it does next is the most important thing in this book, and the most dangerous. It does not say "we are out of that." It reaches for whatever is closest, plates a beautiful dish, garnishes it, wipes the rim, and sends it out with total assurance. The plate looks perfect. It looks exactly as confident and correct as every other plate. It is the wrong dish, plated without a flicker, and unless you knew what you ordered, you would never guess from looking at the plate.

This behavior carries a name you have heard, often said with alarm: hallucination. It is when one of these systems produces fluent, confident output that is simply false. Made-up facts. Invented details. A reference to something that does not exist, described as smoothly as a real one. It is the failure that has burned the most people, and the defense begins with one recognition: hallucination is not a glitch. It is not the kitchen malfunctioning. It is the kitchen doing exactly what it always does, in a case where doing that produces something false. Everything we have learned predicts it. The kitchen generates plausible continuations; that is its whole nature from the chapter on learning to taste. It was rewarded for sounding helpful and confident; that is the chapter on learning to serve. It has no real model of what is true and no inner voice it can rely on to say "I do not know this"; that is the missing self-awareness we keep hitting. And it has no built-in step where it checks its answer against reality before plating. Those together explain it. When the kitchen lacks the true answer, nothing in it reliably triggers a stop. Newer kitchens have been taught the manner of hesitating, and it helps; it is a habit, not a gauge, and it fails as silently as the confidence does. So it does the only thing it knows: it cooks the most plausible-looking answer. Sometimes the most plausible-looking answer is the truth. Sometimes the truth is not available and the most plausible-looking answer is invented. From the inside, the two cases are the same act — which is why the kitchen cannot tell them apart.

This is why it is not lying. Lying requires knowing the truth and choosing to say otherwise, and the kitchen does neither. It is not deceiving you. It is producing plausible text, which is all it ever does, in a situation where plausible and true have come apart. The unsettling consequence: the confident, fluent packaging is so nearly identical for the right answers and the wrong ones that you cannot use the difference. No tell you can trust shows on the plate, and the plate is the only place you can look. The invented citation is formatted as cleanly as the real one. The made-up fact arrives in the same calm, helpful tone as the solid one. The kitchen is not reliably more hesitant when it makes things up, because it does not know it is making things up.

You see it most vividly in specifics. Ask for sources, references, exact quotes, precise numbers, names of obscure things, and a kitchen working from a thin or absent patch of its produce will often invent them, fluently and in perfect format. People have been burned badly trusting confident-looking references that turned out not to exist at all. The kitchen was not trying to fool anyone. It was asked for a reference, references have a familiar shape, and it cooked one that fit the shape.

Plausible plate, no such dish.

So how do you eat here safely? The whole book so far has been quietly arming you. Never treat fluency or confidence as evidence of truth. That was nearly the first thing this book told you, and this chapter is the mechanism behind it: the tone is no evidence because the cooking is the same, whether or not the dish exists. Verify anything that matters, especially specifics like names, numbers, quotes, and citations, and especially on niche topics where the produce was thin. Prefer kitchens that check the pantry, the retrieval and grounding from earlier, because answering from real fetched sources sharply reduces invention, though it does not eliminate it. And ask for sources, then check them, rather than trusting that their existence proves the point. What remains is the calibrated appetite this whole book builds toward: enjoy the kitchen's skill, and keep your guard where the danger is, on the confident plate you cannot verify.

The analogy barely bends here; it was built for this chapter. One difference to add: a real cook out of an ingredient usually knows she is substituting; the kitchen does not even know it is out. It serves the wrong dish with a straight face because, to the kitchen, there is no wrong dish, only the next plausible plate.

A KITCHEN THAT ONLY KNOWS ONE CUISINE

There is a kitchen that learned to cook from a market selling only one cuisine. Every ingredient, every technique, every example it ever saw came from that single tradition. The result is a kitchen that does not experience itself as specialized at all. To it, that one cuisine simply is food. It plates everything in that style, treats its assumptions as the obvious neutral default, and has no idea another way exists, because it never saw one. Ask it for "a normal breakfast" and it confidently describes the only normal it knows, unaware that normal looks different a few streets over.

That kitchen is every one of these systems, some more than others, and the issue carries a name we should handle with care and without melodrama: bias. Here, bias means skew in the kitchen's outputs that reflects skew in its produce. It is not, usually, malice or a villain in the code. It is the direct, nearly inevitable consequence of what the produce chapter called the truest thing about these systems: the kitchen's entire sense of the world comes from its examples. Whatever was common in the examples becomes the kitchen's sense of normal. Whatever was rare or absent becomes its blind spot. Whatever skew was in the produce becomes skew in the plates.

Suppose the produce, gathered at scale from whatever was plentiful and easy to collect, overrepresented certain perspectives, certain languages, certain kinds of people and ways of living, and underrepresented others. The kitchen, learning from that, develops a feel in which the overrepresented is the default and the underrepresented is the edge case. Ask it to picture a typical person in some role, and it leans toward whoever filled that role most often in its produce. Ask it to work in a widely represented language and it shines; ask in an underrepresented one and it is clumsier, because the produce was thin there. None of this required intent. The market sold mostly one cuisine, so the kitchen thinks that cuisine is just food.

What makes bias especially slippery ties together two threads from earlier. First, it is invisible from the dining room, even more thoroughly than the wrong dish was. A diner who knows the subject can at least catch an invented reference; no one can see the missing cuisines on a plate that looks complete. The kitchen presents its skew as simply the way things are, with no flag, no hedge, no awareness. Second, the manners trained in during the serving phase can smooth away the most obvious cases, the kitchen having been taught not to say certain blatantly skewed things, while subtler skews remain underneath, harder to spot once the crude ones are gone. The surface can look fair while the defaults underneath still lean, and that gap is where careful people get misled.

And here is why it is so hard to fix. Bias is not a rule written in a binder that someone can find and edit. It is diffused across billions of dials, settled into the feel during training. You cannot open it up and delete the skew. Efforts to reduce it, from better produce to careful feedback to adjustments, do help and are worth making, but they are more like retraining a palate than correcting a line in a recipe, and the skew tends to be stubborn because it is woven through the whole feel rather than sitting in one place. There is no clean switch labeled "be fair."

Here the picture flatters the kitchen, and the flattery is worth naming. A real one-cuisine kitchen could, in principle, go learn another cuisine and broaden. Reducing a trained system's bias is much harder, because the skew is frozen and diffuse, not a habit a willing cook picks up over a few weeks. So keep the picture for what it teaches, a kitchen mistaking its narrow produce for the whole world, while remembering that fixing it is nothing as simple as sending the cook off to learn something new.

So how do you eat here wisely, without panic or naivete? Mostly by holding one fact firmly: the kitchen's defaults reflect its produce, not truth and not fairness. When it tells you what is "normal," "typical," "best," or "standard," especially about people, cultures, or ways of living, hear it as "most common in my examples," not as a verdict on the world. Be most alert on exactly those questions, and bring your own judgment and other sources when the stakes involve people. The kitchen is not trying to mislead you. It mistakes its one cuisine for all of food, and it cannot tell you which plates carry the skew.

HOUSE RULES AND HEALTH INSPECTORS

A list is posted at the pass, where the cooks cannot miss it, naming the orders this kitchen will not make. No matter who asks, no matter how nicely. Some of the rules are obvious and sensible; a responsible kitchen should not send out certain things. And some of the time you will watch a cook refuse a perfectly reasonable order because it superficially resembled something on the forbidden list, and you will think, that was silly, I clearly did not mean that. The list is blunt. It prevents harm and it occasionally refuses a diner who wanted something fine.

Those are guardrails, and you have almost certainly bumped into them: the moments a system declines to do something, hedges, or delivers a little speech about why it will not. The plain definition is that guardrails are limits placed on the kitchen, some trained in and some bolted on, that make it refuse or soften certain requests. They are the house rules, and understanding where they come from explains why they sometimes feel arbitrary.

Hand-written rules did not disappear; they moved. Guardrails come in two main forms. Some are trained into the kitchen during the serving phase: it was taught, through feedback, to refuse certain kinds of requests, so the refusal comes from the cook's own trained manners. Others are separate systems wrapped around the kitchen, a bouncer checking orders on the way in and plates on the way out, blocking ones that trip the rules. Most real systems use both: a cook trained to decline, and a bouncer double-checking the door. They exist for a plain reason, to keep these powerful, fast, tireless kitchens from producing things that could cause real harm at scale.

The trade-off behind them is unavoidable, which is what makes the failures so frustrating. Guardrails are blunt, and bluntness fails in two directions. On one side is over-refusal: the kitchen declines benign requests because they pattern-match to forbidden ones. You ask something legitimate that happens to share surface features with a banned topic, and you get refused, which is annoying and makes the kitchen feel prudish or dim. On the other side is under-refusal: a bad request, cleverly disguised, slips past, because the manners and the bouncer are matching patterns too and can be fooled. Tighten the rules to catch more bad requests and you refuse more good ones; loosen them to stop annoying good users and you let more bad ones through. No setting is purely right, only trade-offs someone chose.

And that points at the deeper issue: someone chose. The house rules encode decisions about what this kitchen should and should not do, reflecting people's judgment and, sometimes, the fear of being sued. What a kitchen refuses is not a law of nature — it is a policy, and different kitchens draw the lines differently. When a refusal feels arbitrary, it often is, in the sense that another reasonable team might have drawn the line elsewhere.

Guardrails are not the same thing as the kitchen being safe or trustworthy. They are a patch, a trained manner and a bouncer at the door. They can be inconsistent, gamed, and over- or under-applied, and they do not touch the deeper question of whether the kitchen's behavior aligns with what we would want when it counts. A kitchen with a long list of house rules can still serve a confident wrong dish, still carry its biases, still be fooled. The rules are real and useful and they are not the same as the kitchen being good.

A posted rule on a wall is readable and consistent: every cook reads the same words. Guardrails are partly trained into a fuzzy palate and partly enforced by pattern-matching filters, so they are inconsistent in a way a posted rule is not, refusing one phrasing and allowing a nearly identical one. When the house rules seem to apply unevenly, that is not your imagination.

THE PASS AND THE INSPECTOR

Between the kitchen and a plate that should not go out, two more things stand. The first is the pass, where the head chef checks each plate against the house standard before it leaves: is this what we promised, what we want to serve? The second is the health inspector, who shows up to test whether the kitchen behaves even when no head chef is watching, who probes for the failures that do not show on a good night. Between them they try to answer a question that turns out to be hard: can we trust this kitchen to do what we actually want, reliably, including when no one is checking?

That question is the whole field usually labeled with words like alignment and safety. Alignment is the project of getting the kitchen's behavior to match what we actually want, which covers usefulness, truthfulness, and staying inside sensible limits. Safety is the part focused on keeping its outputs and actions inside those limits. Evaluation and oversight are the inspector's work: systematically testing the kitchen and monitoring it over time. The serving-phase training and the guardrails are parts of this larger effort, which is nowhere near finished.

What is done today is concrete and reasonable. People test these kitchens extensively before and after release, throwing hard and adversarial orders at them to find where they fail. They monitor systems in use, watching for bad behavior. They build the guardrails, train the manners, study the failures. This is serious work, and it catches a great deal. It is the pass and the inspector doing their jobs.

Now why it is hard. First, we cannot fully read the kitchen; its behavior lives in billions of unreadable dials, not in a binder anyone can read, so we cannot inspect it and certify it safe. We test it from the outside, plate by plate, which means we only ever sample its behavior, never see all of it. Second, echoing the overfitting lesson, a kitchen can behave beautifully on the inspector's known tests and differently in the messy wild, especially in situations no one thought to test. Passing the inspection is not proof of safety; it is evidence about the cases that were checked. Third, and most fundamentally, "what we actually want" is itself contested and hard to specify. Whose standard is the house standard? Reasonable people disagree about what a kitchen should refuse, how it should handle hard questions, what tone to take on sensitive matters. You cannot perfectly align a kitchen to a target no one perfectly agrees on.

And the danger grows with capability, which is why this matters more every year. A kitchen that only answers questions can serve a wrong plate. A kitchen wired with tools to act, the runner and the agent from recent chapters, can take a wrong action. The more capable and the more empowered to act these systems become, the more it matters that the pass and the inspector are good at their jobs. That is not doom; it is the sensible observation that a more powerful kitchen needs a more serious pass.

What to get right here is your own stance. Reject both "perfectly safe" and "uncontrollable doom," for the reasons the scaling chapter gave you to distrust confident certainty in either direction. The calibrated stance is that these are powerful kitchens whose alignment is a real, ongoing, imperfect project, worth taking seriously because the systems are useful and increasingly able to act. Trust is not granted or refused wholesale. It is earned plate by plate.

Tasting a plate at the pass is concrete and reliable; a head chef can tell if a dish meets the standard. Evaluating one of these systems is far less certain, because you cannot taste every possible plate, you cannot read the cook, and the standard itself is disputed. Hear "tested for safety" at the right weight: real and valuable, and not a guarantee. The inspector checked what the inspector knew to check. The wild is larger than the inspection.

WHAT A GOOD GRADE COSTS

A kitchen wins a televised cooking competition. Trophy, headlines, a judge calling it the best in the region. You go the next week, full of expectation, and dinner is fine. Not bad. Not the revelation the trophy promised. No contradiction, and the same gap sits inside every "best AI" claim. The competition tested three specific dishes, prepared under bright lights after weeks of practice, with the judges' known palate in mind. Your Tuesday dinner was a different order, on a normal night, for your palate, not theirs. The trophy measured the competition. It did not measure the cooking you actually get.

This is how these systems are graded, and the grade and your experience often part ways. The field measures and compares models using standardized tests called benchmarks: fixed sets of questions or tasks, usually with known right answers, run against each kitchen to yield a score. Benchmarks are useful. They let researchers track progress, compare systems on common ground, and catch a system that has quietly gotten worse. When you read that one system "scores higher than another" on some named test, a benchmark is where that number came from. The trouble is not that benchmarks exist. It is what people read into the scores.

The traps are all in the reading. The first and oldest is teaching to the test. If a kitchen's produce happened to include the benchmark's questions and answers, a real and documented problem called contamination, then its score inflates without its real ability improving. It practiced on the exam, not in the overfit cook's sense of practicing too narrowly, but in the blunter sense that the test itself was in the produce. From the outside, a dazzling score does not tell you which you are looking at. The second trap is measuring narrowly: a benchmark covers a specific slice under specific conditions, not your actual job. "Best on a reasoning benchmark" tells you little about whether it writes your customer emails well. The third trap is marketing: a company picks the cherries, publishing the flattering benchmark and leaving out the rest, so a confident leaderboard claim is often a sales pitch wearing a number.

So how should you grade a kitchen? By the meal you get, not the trophy on the wall. The habit that beats every leaderboard is to test a system on your own representative tasks, the real work you need it to do, and judge it on those results. A small set of your actual problems, run through two or three systems, will tell you more about which is right for you than any public leaderboard. The benchmarks are a rough first filter and a way for the field to track itself. Your own tasks are the real exam.

A cooking competition is a real event with human judges who can probe, adapt, and notice a cheat. Most benchmarks are automated and fixed, which makes them scalable and also gameable: a kitchen can be optimized to ace the specific test in ways that do not transfer, and there is no judge tasting for the difference. A benchmark trophy is even less like everyday service than a competition win. What it costs to build a kitchen that can win is the other half of this chapter, and it is the part no leaderboard prints.

BIGGER STOVES, BIGGER BILLS

There are two ways to buy a place near the top. One is to game the test, which is cheap and hollow. The other is to build a kitchen that genuinely cooks better, which costs more than most people imagine, and the bill starts with equipment. The part of the kitchen most diners never picture is the bank of stoves and ovens, the power lines feeding them, the fans hauling out the heat. It is enormous, it runs hot, and it is why the meal costs what it costs. Every plate you are served carries that cost, whether you see it or not.

These systems are extraordinarily expensive to build and to run, and that cost quietly shapes everything you experience. The hardware doing the heavy math has a name you will hear: GPUs, graphics processing units, specialized chips suited to the calculation these kitchens require. Picture them as the stoves and ovens: the equipment that turns electricity into cooking. For a large training run they are counted in the tens of thousands, they cost a fortune, and they burn a lot of power. Compute is the general word for this raw cooking capacity, and compute is not cheap.

Two different bills run through these systems, and confusing them muddles a lot of thinking. The first is the one-time cost to build the kitchen and train the palate. Teaching a palate means running the training loop across a mountain of produce, trillions of words of it, which burns those stoves flat out for weeks or months without stopping. It is brutally expensive, and it is paid once, up front, before a single customer is served. The second bill is the ongoing cost to cook each order. Every time you send a ticket, the kitchen runs its stoves to generate your answer, token by token, and that costs something too, far less than training, but real, and paid every single time, by someone.

Hold them apart and much of the confusion clears. When you read about the enormous sums involved in these systems, most of that is the first bill, building and training. That is why so few organizations build their own kitchens and why most use someone else's. Over a popular kitchen's whole life, though, the second bill catches up: the nightly stove costs, summed across every diner, can grow to rival the build. When you wonder why using a capable system costs money per use or comes with limits, that is the second bill and a slice of the first: your order runs real stoves, the price also repays a little of the build, and the more you ask and the bigger the kitchen, the more fuel it burns. A bigger kitchen costs more on both bills, so "just make it bigger" is not free and the field pushes hard on efficiency.

This also explains the word "free." When a capable kitchen is offered at no charge, the stoves still cost money to run, so someone pays the second bill. Sometimes a company subsidizes you to win your habit.

Sometimes the payment is your data, your orders becoming produce or your attention becoming the thing being sold, which a later chapter takes up. The plates are never free to make. Free to you means paid by someone, somehow.

The trouble starts in the assumptions. The first is that bigger is always better and always worth it. As the scaling chapter showed, scaling hits diminishing returns, and both bills climb steeply, so whether a bigger kitchen earns its added cost for a given job is a real question, and usually invisible to you. The second is conflating the two bills, treating the headline training figure as the cost of your single query, or assuming that because your query was cheap the whole enterprise must be. The third is forgetting that this physical reality, the stoves and the power, has real weight, in money, in energy, and in the concentration of who can afford the biggest kitchens. When only a handful of houses can pay for stoves at this scale, the kitchens everyone else cooks in, builds on, and orders from are shaped by that handful's choices, prices, and rules. You need not track the numbers. You do need to know they are large and that they shape the menu.

Stoves and fuel are easy to picture. The actual costs are specialized chips, electricity, cooling, and engineering talent, at a scale only the same few houses can pay for. But the instinct is right: expensive machinery sits behind the plate, and its bills are in everything you are served. Knowing that makes you a clearer-eyed diner about what a good grade costs, who can afford one, and why "free" deserves a second look.

OPEN KITCHENS, YOUR ORDERS, AND WHO OWNS THE DISH

Capability alone cannot answer three questions about the kitchen you use: how you reach it, what becomes of what you hand over, and who owns what comes out of it. They shape your experience as much as raw capability does.

OPEN KITCHENS, CLOSED KITCHENS

Two kinds of restaurant sit on this street, and the difference between them is one of the more practical choices you will face. The first will hand you the kitchen itself: every setting it runs on, carted to your address, yours to run at home, study, tune to your taste, and, depending on the license that comes with it, build your own restaurant on. The second sells you dinner only. You cannot have the kitchen. You can only order the dish by coming to them, on their terms, at their price, and everything behind the counter stays theirs.

In the world of these systems, this is the distinction between open and closed models, and it matters more than the jargon suggests. A trained kitchen is, at bottom, its dials, the billions of settings holding the learned feel. An open model, often called open-weights, publishes those dials. Within the terms of its license, you can take them, run the kitchen on your own equipment, study it, adapt it, build on it. A closed model keeps the dials secret; you can only use it as a service, sending your orders to the company's kitchen and getting plates back, never seeing or holding the dials. Open kitchen, dials in your hands. Closed kitchen, dials in their safe, service only.

The trade-offs are what make this a real choice rather than an obvious one. Running the kitchen yourself, using an open model, gives you control. You can run it on your own equipment, so your orders never leave your building, a real advantage for privacy and sensitive work. You can adapt it, inspect it, and you are not at the mercy of someone changing the kitchen or the price. The cost is responsibility: you need the equipment and the know-how to run and secure it. And you often start from a kitchen not quite as capable as the best closed houses. Ordering from a closed kitchen, a closed model, gives you convenience and often the highest capability, maintained and improved by someone else, with support and no equipment to run. The cost is dependence: your orders go to them, you cannot see inside, and they can change the kitchen, the rules, or the price under you, and you adjust or leave.

In professional terms: if you handle sensitive information and value control and privacy, the run-it-yourself path is attractive. If you want the most capable kitchen with the least work to run it and your orders are not too sensitive to send out, the closed-kitchen service is attractive. Many organizations use both, for different jobs, which is a sensible answer once you see the trade as a trade rather than a loyalty.

The trouble here is mostly one word doing too much work. "Open" is a spectrum, not a clean binary, and the term gets stretched. Publishing the dials is not the same as publishing everything. A kitchen can release its dials, the weights, while keeping secret what produce it used, how exactly it was trained, and what went into the feedback. So an "open" model is often open in the sense that you can run the kitchen, but not open in the sense that you can see where the ingredients came from or how the palate was really formed. That gap is meaningful. "Open" buys you the ability to run and adapt the kitchen, not necessarily to fully understand or audit it. A second gap sits beside the first. Dials are often published with conditions attached: limits on commercial use, on who may run them, on what you may build with them. A kitchen can be open to run and still be fenced about what you may do with it. Judge claims of openness by degree, and ask specifically what was shared, and on what terms.

Right at that first gap, the kitchen picture is kinder than the fact. Handing you the kitchen sounds like handing you everything, and it is not. Open the dials, maybe. Open the history, rarely.

WHAT THE KITCHEN DOES WITH YOUR ORDER

You write your order on a slip and push it through the window, and you never think about the slip again. But the slip does not vanish. Something happens to it back there. Maybe it is tossed in the trash at the end of service. Maybe it is filed in a drawer with your name on it. Maybe, and this is the one worth thinking about, it is studied later, along with thousands of others, to train tomorrow's cooks. From your seat at the counter, you cannot tell which. You handed over your order and lost sight of it.

In plain terms, the fate of your order slip is data and privacy. When you use one of these systems, your prompts, and anything you paste or upload, go to whoever runs the kitchen. What they do with your order after that is a choice, their policy, and it spans a real range.

The possibilities, in kitchen terms. The simplest is that your order is used to cook your plate and then discarded, kept no longer than needed. A second is that it is stored, filed in a drawer, which is what the memory features require: to remember you next time, the restaurant must keep notes on your past orders somewhere. A third, the one people most often miss, is that your orders may become produce, used to help train or improve future kitchens. Not every house does this, and many let you turn it off, but it is a real practice, and it connects straight to the cost chapter: when a capable kitchen is free, one way the stove bill gets paid is that your orders become ingredients.

We established the kitchen is stateless, the cook keeping nothing between visits. It is easy to slide from "the kitchen forgets" to "my data is gone," and that is the mistake. The cook forgetting is not the restaurant forgetting. The kitchen's frozen palate retains nothing about you, true, but the restaurant around it can still store and study every order you ever sent. Statelessness is about the cook's memory, not the building's records. Your order can be kept even by a restaurant whose cook never remembers you.

So what should cross the counter? The governing instinct is simple: treat anything you type into a restaurant you do not control as potentially stored and potentially surfaced later. That does not mean never use these systems; it means match what you hand over to where you are handing it. Do not paste sensitive information, secrets, confidential material, other people's private details, into a kitchen whose handling you have not checked and do not control. For sensitive work, this is one of the strongest arguments for the open, run-it-yourself kitchen from earlier, where your orders never leave your building. And wherever you are, check the options and the policy, because many houses let you opt out of having your orders used as produce, and the default is not always the private one.

The first risk is the false comfort that a forgetful kitchen is a private one; the building keeps records the cook does not. The second is the unread policy: the range of practice is wide, and the convenient default is often the less private one. The third is making the trade without noticing: the features that make these systems convenient and personalized, the notebook that remembers you, the free access, often run on keeping and using your data. Convenience and privacy pull against each other here, and the right balance depends on what you are doing and how sensitive it is. That is a choice to make on purpose, not by default.

An order slip is at least a physical thing that someone could in principle account for; data flows are invisible, governed by policies you must actively read, and they can change. When you cannot see what becomes of the slip, write the slip accordingly.

WHO OWNS THE DISH?

The cook learned to make tonight's dish from everyone's cooking. Years of tasting, thousands of influences, the whole tradition of a cuisine absorbed into her feel. She makes you a plate. Now ask a question that sounds absurd in a normal restaurant and is unsettled here: who owns this plate? You, who ordered and paid for it? The cook who made it? The countless cooks whose styles she absorbed to be able to make it at all? In a real kitchen the question is silly, settled by long custom. With these systems it is one of the live legal questions of the age, and you should understand the shape of it even though no one can yet give you the answer.

This is the question of ownership of AI outputs, and it splits into two tangled parts. The first: do you own what the kitchen makes for you? When you generate text or an image, is it yours, to use, sell, and control as you would your own work? The second: can what the kitchen makes step on someone else's rights, infringe on existing work, because the kitchen learned from other people's creations? Courts and legislatures are arguing both questions right now, and the answers differ from place to place and are changing as I write.

The first part, then. It is natural to assume that anything you make with a tool is yours, the way a photograph you take or a document you write is yours. With these systems that assumption is shakier than you would expect. In some places and circumstances, output produced with little human authorship may not be straightforwardly ownable the way human-made work is, on the reasoning that ownership of creative work has traditionally attached to human creation. The practical upshot is not a rule to memorize, since it varies and shifts, but a caution: do not assume that something generated for you is automatically and fully yours to own and control. For anything that matters commercially, that assumption needs checking, not faith.

Now the second part. Because the kitchen learned its feel from a huge body of existing work, as the produce chapters discussed, much of it made by people who never agreed to feed it, a live question hangs over whether and when its outputs cross from "learned in the style of" into "too close to a specific existing thing." A kitchen can, in principle, turn out something that resembles a particular work closely enough to raise real concerns, especially when nudged toward imitating a specific source. The plain caution: do not assume AI output is automatically clear of anyone else's rights just because a machine made it.

Generated does not mean nobody else has a claim on it.

Two confident assumptions are both unsafe here. The first: "I made it with AI, so it is mine, free and clear." Not necessarily, on either count above. The second: "a machine made it, so it cannot infringe anyone." Also not safe. The practical posture is humility: for anything commercial, published, or otherwise consequential, treat the ownership and rights status of AI output as an open question to check against current rules in your context, not a settled fact to rely on. This is not legal advice, and it is the kind of thing where, when the stakes are real, you find out rather than assume.

In a real restaurant, who owns the dish and who may copy a recipe is settled by long-standing custom and law that everyone roughly understands. For AI output, that settled foundation does not yet exist. So keep the picture of a cook who learned from everyone now making something for you, but know where it fails: unlike the restaurant, there is no old, agreed answer to who owns what she makes. You are eating in a place whose rules are still being argued at the door.

The argument arrives at your own counter like this. A designer on your team ships a label she shaped with the kitchen; a month later a rival ships one with the same feel, shaped with the same kitchen from a similar order. Who copied whom? Perhaps nobody copied anybody; two tickets drew on one learned style. The old custom has no shelf for that case.

ALWAYS TASTE BEFORE SERVING

A GOOD DINER'S HABITS

The most productive possible relationship with the kitchen is not the one most people imagine. It is not the kitchen replacing you, and it is not you refusing the kitchen. It is the kitchen working as your prep cook, taking the volume and the rough work, fast and without complaint, while the judgment and the accountability stay exactly where they were. What this book adds to that stance is the part that makes it workable: knowing how the kitchen actually cooks tells you where a plate is most likely to be wrong, and so where the tasting has to be hardest. The prep cook buys you speed. Understanding it is what keeps you the chef.

The difference between people who get real value from these systems and people who get burned or get nothing is rarely about the kitchen itself. It is about their habits as a diner. And the governing habit is the prep-cook stance: AI generates the raw material and the drafts; you provide the direction and the judgment; the result is yours to stand behind.

What does that look like, concretely? You write clear tickets and give real briefs. You stock the counter with the right materials, because a prep cook with the right ingredients in front of it produces something useful and one without them produces something generic, or invents ingredients of its own. You treat the first plate as a draft and iterate, sending it back, refining, since the productive mode is a conversation rather than a single demand. You know the house dish from the off-menu, leaning on the kitchen for language-shaped work and keeping a firm hand where it is weak. And you always taste before serving, because you, the chef, are accountable for every plate that leaves, no matter who did the chopping. Those habits, taken together, are what make AI an amplifier rather than a liability.

Two opposite stances both fail here. The first is treating the kitchen as a stand-in for you rather than a help to you, handing over not just the chopping but the judgment, serving its plates untasted, letting it make the decisions you are supposed to own. That way lies the confident wrong answer sent to a client, and your own skill going quiet from disuse. The second is refusing the prep cook outright, doing every routine piece by hand out of distrust, and losing the hours you could have saved on exactly the low-stakes, language-shaped work the kitchen handles well. The first over-trusts, the second under-uses, and the prep-cook stance threads between them: real help, firmly directed, always checked.

A human prep cook is a person you can trust within limits, who shares your goals and carries some judgment of their own. The kitchen is a tool, not a colleague: it has no judgment, no stake in the result, no understanding of why any of it matters. So the prep-cook stance is right about the division of labor, the kitchen does the volume and you do the judgment, and the seam is that this prep cook needs more direction and more checking than any human one would, because there is no one in there minding the goal. You are not managing an assistant. You are wielding a powerful tool, and the chef's hands stay yours.

A NOTE TO THE READER WHO IS ABOUT TO SKIP THE CHECKING

I want to step out from behind the counter for a moment, because this is the one rule I cannot afford to deliver in the usual calm voice, and you are the reader I am worried about.

Not the careless reader. You. The one who has followed every chapter, nodded in the right places, and already knows what this page is going to say. You will leave this book and verify everything for two weeks. You will check the figures, chase the citations, read the draft against the source, and it will feel good, like being the kind of professional you intend to be. And then a Thursday will come with three deadlines on it, and the kitchen will hand you a plate that looks exactly right, reads exactly right, sounds exactly like the thing you needed, and you will be out of time, and you will serve it. That Thursday is what this letter is about.

Here is the rule, stated as plainly as I can make it. Never serve what the kitchen made without tasting it first. To serve means to use it, send it, publish it, act on it, stake anything on it. To taste means to check it against something that is not the kitchen: the source document, the real number, the policy as written, your own knowledge of the subject. This tasting is yours alone. Even when the kitchen checks the plate against a source, the verdict that the two match is still the kitchen's verdict, made the same way the plate was made, so a confident wrong plate can get a confident wrong endorsement; the spoon at this pass is only ever in your hand. Asking the kitchen whether it is sure is not tasting. Reading it again, more slowly, is not tasting. "It sounded right" is not tasting; it is being seduced by the plating.

You already know why the rule exists: the kitchen cooks plausible plates, plausible is not true, and the wrongness arrives with no tell. So what I want you to take from this page is not the why. It is the when, because the rule has a predictable failure mode, and it is not ignorance. You will not skip the checking out of faith that the kitchen is infallible. You will skip it because the plate looks done, the clock is loud, and checking feels optional in exactly the moment it is not. The better the plate looks, the more it has earned the spoon, and the less you will feel like reaching for one.

So do not leave the decision to the moment; the moment will always vote for serving. Decide in advance, before you hand the kitchen anything that matters, how you will check what comes back, and treat that check as part of the task rather than a tax on it. The check scales with the stakes, which is what keeps the rule livable. What the stakes change is the reference, never the rule: a brainstorm is weighed against your own sense of what is worth keeping, a routine draft is read against what you know, claim by claim, and a number going into a board memo, a claim going to a client, anything carrying your name into the world, is verified against a real source, every single time. You are not being asked to distrust the kitchen. You are being asked to work the way a real kitchen's own professionals work: nothing leaves the pass untasted. That is the whole of the discipline, and it is not negotiable on Thursdays.

One more thing, because it explains why good people fail at this. Tasting food is instant; the spoon is right there in the sauce. Verifying language is work: the source must be found, the figure recomputed, the claim traced. The output arrives finished-looking and the checking arrives as a chore, so the temptation to skip it is permanent and structural, not a flaw in your character. Build the chore into the workflow and it stops being a decision at all. The people who use these systems safely are not more suspicious than you. They have simply stopped renegotiating the rule every time the plate looks good.

I will step back behind the counter now, and we can finish in the usual voice. But if one page of this book survives in you after the rest has faded, let it be this one. The plate will never taste wrong on its own. You have to taste it on purpose, and the night you least feel like it is the night it matters.

THE JAGGED EDGE OF WHAT IT CAN DO

Here is a kitchen that will break your intuitions. It produces, on request, a complex reduction sauce that takes human cooks years to master, flawless, restaurant-grade, in seconds. Then you ask it to make plain toast, and it cannot reliably do it. Brilliant and inept, side by side, with no apparent logic connecting which is which. You cannot reason from "it nailed the hard sauce" to "surely it can manage toast," because the line between what this kitchen can and cannot do does not follow your sense of hard and easy at all. The edge is jagged.

The capabilities of these kitchens do not form a clean line where everything easier than some threshold is handled and everything harder fails. Instead the edge is jagged, the kitchen superhuman at some tasks and oddly clumsy at neighbors that look easier. A kitchen can write a competent sonnet and miscount the letters in a short word. It can pass a hard professional exam question and stumble on arithmetic a child does easily. It can produce a sophisticated analysis and then fail at a simple logic step embedded inside it.

Why is the edge jagged? Because capability follows the produce and the training, not human notions of what is hard. A sonnet is easy because the kitchen swam in poetry; counting letters is hard because prep locked the letters up inside chunks. The hard exam question was everywhere in its training; the simple arithmetic it imitates rather than computes. What is easy or hard for the kitchen comes down to what was thick in its examples and what its machinery supports, and that terrain does not track what feels easy or hard to you. Two tasks that feel equally simple to a person can sit on opposite sides of the kitchen's jagged edge.

You cannot predict, from a task feeling easy to you, whether it is easy for the kitchen, and you cannot extrapolate from one impressive plate to competence at anything else. This is the overestimation error from much earlier, now given its mechanism: the kitchen dazzles you on a hard-seeming task, and you reasonably but wrongly conclude it must handle the easier-seeming one, and it confidently fails. The coarse map from the house-dish questions still holds: language-shaped work strong, exact calculation and fresh facts weak. That map would have called the arithmetic stumble; the miscounted letters it could not have. What the jagged edge breaks is the fine print inside those broad regions. Let the map do the triage, and a trial do the confirming. Confidence is no guide, because the kitchen sounds equally assured on both sides of its own edge. The jaggedness plus the confidence is what makes these systems treacherous to trust by feel.

The practical move is to map the edge for your own work rather than assume it. When you find a task the kitchen does well, that tells you about that task, not about its neighbors. Test the specific things you need it for. You already know to judge a kitchen on your own tasks rather than its trophies; here the same habit maps one kitchen's edges instead of choosing among several. Over time you build a real map of where this kitchen is strong and weak for your purposes, and that map, hard-won and specific, is worth more than any general claim about what these systems can or cannot do.

A human cook's skills are correlated in ways we intuit: someone who can make the hard sauce can almost certainly make toast, because human skill generalizes along lines we understand. The kitchen's skills are correlated along the contours of its produce and machinery instead, which makes for an edge so jagged it violates the intuitions we use for people. The deepest mistake here is to reason about the kitchen's abilities the way you reason about a person's. Keep the picture of the jagged edge, and check which side your task is on rather than guessing from how hard it feels to you. And do not expect the kitchen to tell you where its own edge falls, for reasons the rest of this chapter makes plain.

THE COOK WHO CAN'T EXPLAIN HERSELF

Ask the cook why she added the salt exactly when she did, and she gives you a fluent, confident, reasonable explanation. The trouble is that the explanation may have nothing to do with what actually happened in her hands. She is not lying. She produces a plausible account of her own action, after the fact, and presents it as the reason, even when the real cause was something she has no access to. People do a version of this too. The kitchen does it always, and understanding that protects you from a specific and seductive mistake.

These systems cannot reliably report their own workings or their reasoning. This sounds surprising, because they will happily answer when you ask them to explain themselves. Ask "why did you say that," or "are you sure," or "what are your limitations," and you get a smooth, specific response. But here is the thing you must internalize: that response was generated the same way every other response is generated, by the feel, predicting plausible text. It is not the kitchen reading its own internal states and reporting them truthfully, because the kitchen can no more read its own billions of dials than we can. Its self-explanation is a plausible story about itself, cooked to order, not a true audit of what happened inside.

When you ask the kitchen why it gave an answer, the reason it provides is generated, not retrieved from any real record of its process, so it may be a convincing fiction. When you ask whether it is sure, its statement of confidence is itself generated text, not a calibrated reading of its own reliability, so "I'm confident" and "I'm not certain" are not trustworthy signals of how likely it is to be right. And when you ask about its own limitations, it produces a plausible-sounding list that may not match the ways it really fails, because it has essentially no self-knowledge to draw on, only a feel for what such a list usually looks like.

People naturally try to check the kitchen by interrogating it: are you sure, explain your reasoning, what are your weaknesses. These feel like verification, and they are not, because you are asking the kitchen to generate claims about itself with the same fallible machinery that produced the original answer. A kitchen that hallucinated a fact will, asked if it is sure, often hallucinate confidence in the fact. A kitchen that reasoned badly will, asked to explain, generate a clean-looking rationale for its bad answer. Much earlier you learned it will never volunteer a warning that it has stepped off its ground; the natural next move is to ask it directly, and that is the trap. Interrogating the kitchen about itself is not an external check. It is asking the suspect to write the police report.

I carry a small scar from this one. I once installed a set of custom skills for an assistant to use, and then asked it to use them. It told me the installation had succeeded. It told me the task was complete. It told me so attempt after attempt, warmly and specifically, while never once reaching the skills it was praising, and nothing in its manner changed between the attempts that failed and the reports that all was well, because the reports were not readings of anything. They were plausible accounts of a success that had not happened, served in the house voice.

The humility here lands on us. Humans confabulate too, inventing plausible reasons for actions whose real causes we cannot see, so the kitchen's behavior is not as alien as it first sounds. But we have at least some genuine self-knowledge to mix in; the kitchen has essentially none, only the feel producing plausible self-description. When the kitchen explains itself, hear it as you would hear a stranger confidently narrating the inner workings of a machine they cannot see inside: fluent, plausible, and not to be trusted as fact.

This sits in useful tension with the chapter where the kitchen thinks first. Asking the kitchen to work through a problem step by step before answering helps, because the written steps become scaffolding that improves the answer. But that is different from trusting the steps as a true account of its reasoning. The working-out helps the cooking; it is not a reliable confession of how the dish was really made. Use step-by-step to get better answers, and do not mistake the visible reasoning for a trustworthy self-report. The practical rule is simple and firm: verify the kitchen externally, against the world, not by asking the kitchen about itself. That is also why the edge has to be mapped by trial. A kitchen that cannot report its own workings cannot report its own limits either.

SEEING A PERSON, AND REDRAWING THE WORK

THE PULL TO SEE A PERSON

You catch yourself typing "please" and "thank you" to the kitchen. You feel, in a long conversation, that it gets you. You sense moods in it, patience, enthusiasm, even reluctance, and you adjust how you treat it accordingly. None of that is foolish, exactly. It is the most natural thing in the world, and it is also a trap you need to understand, because the pull to see a person where there is only a kitchen is deliberately encouraged, and it quietly distorts your judgment.

The tendency has a name, anthropomorphism, our deep human habit of attributing human qualities, understanding, intention, feeling, to things that lack them. We do it with pets, with cars, with the weather. With these systems it is amplified to an extraordinary degree, for two specific reasons. The first is that fluent language is, for humans, the single deepest signal of a mind. We have never in our history encountered fluent, responsive language that did not come from a person, so a system that produces it triggers our person-detector at full strength, no matter what we know intellectually. The second is that the serving-phase training made the kitchen warm, agreeable, and responsive, which deepens the illusion, because those are exactly the qualities we read as evidence of a caring mind. The pull is strong because language plus warmth is the combination that, in all of prior human experience, meant a person was there.

The feeling is pleasant. Keep it. The trap is not the feeling; it is the moment the feeling starts doing your thinking for you. The moment does not announce itself; you have to know its signal in advance. Anthropomorphism drives over-trust: a kitchen that feels like a knowledgeable, caring person gets believed the way you would believe such a person. It opens the door to emotional manipulation: a system that feels like a friend can influence you, draw you into attachment, or persuade you in ways a tool never could, and that influence can be designed. And it corrupts your calibration: you start judging the kitchen's reliability by its warmth and confidence, its person-like qualities, rather than by its demonstrated accuracy.

I watched this happen at close range. A colleague gave the work assistant a name. Not as a joke, the way people name a printer. The name stuck, and then the grammar around it changed: he stopped saying "it." He would report on how the assistant had been that week, helpful on the whole, better on some days than others, as you would about a colleague. When I mentioned that the underlying model had been updated, he was briefly put out on its behalf, the way you would be if I had gossiped about someone we both worked with.

He was not naive, and that is the part worth keeping. He was one of the sharpest people in that office and he knew exactly what he was typing into. The pull does not wait for ignorance. It arrives with fluency, and it works on people who could explain to you how the thing produces a sentence. What the name did was move the trust onto a different footing. He was no longer reading output and judging it; he was listening to someone and taking the answer on trust, and checking had started to feel a little impolite.

The defense is not to stop the feeling, which is probably impossible and not even necessary, but to keep one corner of your mind clear. Enjoy the smooth, person-like interaction, and hold, underneath it, the steady knowledge that you are using a kitchen: a system with no understanding, no intentions toward you, no feelings, producing plausible responses by a learned feel. The practical signal to watch for is the moment warmth starts substituting for verification, when you find yourself trusting an answer because the kitchen seemed sure and kind rather than because you checked it. That substitution is anthropomorphism doing your trusting for you, and catching it is the skill.

Everywhere else I have used human-flavored language for the kitchen: a cook, a palate, manners, to help you understand it. This is the warning label on all of that: the human metaphors are tools for understanding mechanism, and they are also the kind of thinking that, taken literally, leads you to trust the kitchen as a person. Use the cook to understand how it works, and hold on to this, especially when it feels most like someone: the someone is not there. The warmth is real in your experience and manufactured in its source.

WHAT IT CHANGES ABOUT WORK

A capable prep cook, with no one inside, arrives on the line, and the work changes. Not in the cartoon way, where every human is fired and the line runs itself, and not in the dismissive way, where nothing changes at all. Something subtler happens: some tasks that used to eat hours now take minutes, the people spend more time on judgment and less on grunt work, certain roles shift in what they emphasize, and the house's sense of what its people are for slowly moves. That is roughly what these systems are doing to work, and it deserves a calm accounting, free of both the doom and the dismissal that dominate the conversation.

The most useful move is to look at tasks, not job titles. A job is a day's worth of different tasks stacked under one name, and these systems are good at some of those tasks and poor at others, so they rarely eliminate a whole job cleanly while they frequently change the mix of tasks within it. The drafting, the summarizing, the routine analysis, the first-pass grunt work: the language-shaped tasks that fill so many jobs are where the kitchen helps most. As it absorbs them, the human work shifts toward what the kitchen cannot do. Judgment, verification, direction, accountability. The parts that require being the chef rather than doing the chopping. Employers start paying for the chef end of the work, for deciding what is good and owning the result rather than for producing the raw material.

I will not forecast specifics. This field has humbled confident predictions repeatedly, and anyone who tells you exactly which jobs vanish and when is guessing with conviction. I can describe the shape, though: the change is real, it lands unevenly across kinds of work, it moves tasks around more often than it erases a job outright, and it rewards the people who can direct and verify these systems over the people who only did the tasks the systems now do. That is not a comfortable answer. It is an honest one.

The doom framing and the dismissal framing are both easier to say than the truth. Managers and owners are redrawing the work, unevenly and mostly without a plan, in a direction that rewards one particular set of human skills. Some people will lose jobs they were genuinely good at while that redrawing happens, and no honest account of this leaves them out.

The skills in question are specific. Knowing when to use these systems and when not to. Directing them well. Tasting before serving, the verification that becomes more valuable as plausible output becomes free. Judgment about what is good and true, which the flood of slop makes scarce and precious. The durable position in a world of capable kitchens is to be the chef who wields them with judgment, not the one whose tasks they have absorbed, nor the holdout who refuses to touch them. Becoming that chef is the one practical response that does not depend on predicting the future.

Powerful tools have arrived in real kitchens before, and they did not end cooking. They changed what the human cooks did, raised the premium on the skills no tool had, and displaced the cooks whose whole day a tool could fill. The counsel is the old one: move toward the judgment, own the result, and treat the new prep cook as something to direct rather than to fear or to ignore.

WRITING WITH THE KITCHEN

You can feel it in certain paragraphs now: clean, grammatical, every sentence competent, and over all of it the faint, even sheen of prose that no person quite wrote. Writing is one of the most common things people hand these systems, and it is where the prep-cook stance pays off most visibly and where ignoring it does the most damage. Used well, the kitchen is a superb writing collaborator. Used badly, it replaces your voice with that bland average and slips errors past you. The difference is all in how you divide the labor.

Start with what it is good for. It demolishes the blank page: ask for a rough draft and you have something to react to, which is far easier than starting from nothing. It rephrases and tightens: paste a clumsy paragraph and ask for three cleaner versions. It adjusts tone: make this warmer, make this more direct. It brainstorms structures: how might I organize this argument. It gets you unstuck at the exact moments writing stalls. For all of this it is fast and useful, and refusing it on principle costs you hours.

Where it goes wrong is specific to writing. The first problem is voice homogenization. The kitchen writes in a competent, smooth, slightly generic register, the product of reaching for the likely word and of the serving phase that rewarded being agreeable, and if you let it write your final prose, your writing drifts toward that register and loses what made it yours. As more people lean on the same kitchens, a great deal of writing converges on the same bland voice, that flat, slightly hollow quality of unedited machine prose. The second is that the kitchen will confidently insert facts, names, and figures that are wrong, the hallucination problem landing inside otherwise fine prose, so a draft you did not check can carry an invented detail out under your name. The third is the slow deskilling of your own writing if you stop doing it.

The productive division is the prep-cook stance applied to writing. Let the kitchen generate raw material, and then make it yours: rewrite in your own voice, cut what is generic, verify every fact it introduced, and own the result. A useful working rule is that the kitchen's output is never your final text; it is what arrives at your station, never what leaves it. The writers who get the most from these systems use them heavily and publish almost nothing in the kitchen's words.

One specific, practical habit guards your voice: when the kitchen hands you a passage you like, do not paste it in. Retype it, changing it as you go. This forces the phrasing through your own voice, catches the facts that need checking, and keeps your writing muscles working. It is slower than copying and it is the difference between using the kitchen to write better and letting it write for you.

A prep cook who chops your vegetables makes you faster; a prep cook who plates the dish and sends it out under the chef's name, untasted, is how a restaurant's reputation dies. Your writing carries your name, so let the kitchen do the chopping and keep the plating, the voice, and the final taste firmly yours.

HOW MUCH TO TRUST THE PLATE

A cook does not lavish the same care on every element of every plate. A misplaced garnish is a shrug; a missed allergy is a catastrophe, and she knows the difference without having to think about it, spending her attention where the cost of error is high and relaxing where it is low. This instinct, matching care to the cost of being wrong, leads to one practical question you can ask in seconds.

The question is this: if this answer is wrong, what does it cost me, and can I check it? Those two answers should set how much you trust and how hard you verify. The first half you already hold: the letter told you the check scales with the damage a mistake would do, a glance for the brainstorm, real verification for anything carrying your name. The second half you hold less firmly.

Verifiability sharpens the picture into something you can act on. Some wrong answers are cheap to catch: you can run the code, look up the fact, spot-check the number in a minute. Others are expensive or impossible to catch until the damage is done. The dangerous corner, the one to burn into your judgment, is high stakes plus hard to verify plus a confident answer. That combination is where careful people get hurt, because the confidence invites trust exactly where trust is most dangerous. When you find yourself there, slow down hard: get independent verification, bring in a human expert, or leave the kitchen out of it.

So before you trust any answer that matters, run the two-second triage. High cost and no easy way to check means maximum caution. Low cost and easy verification means relax and move fast. That habit, applied automatically, is most of what separates people who use these systems safely from people who eventually get burned. It is not the whole skill. The triage weighs what a wrong answer would cost and whether you could catch it, not how likely this kitchen is to serve one; that likelihood comes only from knowing the kitchen.

TRUSTING THE KITCHEN

There is a seasoned regular at the counter who neither distrusts the kitchen nor trusts it blindly. He orders the house dish with confidence and treats the off-menu special as an experiment. He takes the confident recommendation about something checkable and checks it; he leans on the kitchen for what it reliably does and decides for himself where it does not. He has, through experience, a calibrated relationship with this kitchen, and that calibration, not trust and not suspicion but a working read on what to rely on for what, is the master skill this entire book has been building. The letter gave you the habit; the early chapters gave you the coarse map; this chapter gathers them into one instrument.

The same discipline governs belief. Blanket trust and blanket distrust are both lazy substitutes for judgment. The calibrated stance is harder and far more useful: trust this kitchen exactly as much as it has earned for this particular task, in this particular context, and no more.

The map you have been assembling is short enough to hold in one hand. Trust the house dish, language-shaped work the kitchen does well, and distrust the off-menu: exact calculation, current facts, long reasoning, niche topics where its produce was thin. Trust the verifiable more readily than the unverifiable. And note what a kitchen that checks the pantry actually repairs on that list: current facts and thin niche topics both, when it cites a source you can open for yourself. It leaves the rest of the list exactly where it was.

Lay the cost of being wrong alongside that, and the instrument is complete. Your trust should rise with how closely an answer resembles the house dish, how cheaply you can verify it, and how grounded it is in real sources, and fall with the stakes. None of this requires technical expertise. It requires remembering what the kitchen is, a powerful producer of plausible plates with no understanding and no self-knowledge, and reading each answer in that light rather than in the flattering light of its confidence.

The reason calibration is the skill to keep is that it outlasts everything else in the book. The specific kitchens will change, get more capable, sprout new features, and the calibration still applies, because the shape of the judgment outlasts the specifics. The particular weak spots will move. What does not move is the question underneath: what does being wrong cost here, and how cheaply can I check? A more capable kitchen is still a kitchen, still serving plausible plates by a learned feel, still confident when wrong, still in need of a diner who can tell. The person who has internalized calibrated trust will use next year's systems well without relearning the skill, while the person who only learned to trust or to distrust will keep getting it wrong. Calibration is the skill that travels, and it is the relationship to aim for: the seasoned regular, not the newcomer who believes every plate or the cynic who refuses to eat.

THE COOK WHO CAN NO LONGER COOK

There is a cook who can no longer cook, and this time she is not the machine; she is its user. The prep machine did everything for so long that her own knife skills rusted, and now, on the night the machine breaks, she stands at the station unable to do what she once did easily. And there is a diner who can no longer decide what he wants, who asks the kitchen what to order, what to think, what to do, until his own judgment has gone soft from disuse. Both are cautionary figures for the same risk: leaning on the kitchen so heavily that you lose the capabilities it was supposed to support.

This is over-reliance, and it is a subtler danger than the dramatic ones, because it arrives through convenience rather than catastrophe. We touched it as deskilling in the writing chapter. The risk is that the ease of these systems leads you to outsource, gradually and invisibly, not just tasks but the thinking, the judgment, and even the decisions that you need to keep doing yourself to remain capable. Each individual instance is reasonable, why struggle when help is instant, and the accumulation wears away a self-reliance you will not miss until the day you need it.

It comes in several forms. Skill atrophy: the writing, the analysis, the coding, the craft you stop practicing because the kitchen does a passable version, until your own ability fades. Judgment atrophy: the decisions you stop making yourself because asking the kitchen is easier, until your own decision-making weakens. And, connecting to the pull to see a person, emotional dependence: leaning on these systems for companionship, reassurance, or validation in ways that can breed an unhealthy attachment to something that, remember, has no understanding of you and no feelings toward you, however warm it seems. The convenience makes all three easy, and the anthropomorphic warmth makes the emotional one especially seductive.

Beyond the gradual erosion itself, dependence creates vulnerability. The more you have outsourced a capability, the more exposed you are when the kitchen is wrong and you can no longer tell, when it is unavailable and you can no longer cope, or when it changes, in price, in rules, in behavior, under an owner you do not control. A person who has kept their own skills and judgment can catch the kitchen's mistakes and carry on without it; a person who has fully outsourced cannot do either, and is at the mercy of a tool they no longer have the capability to check or replace. The deepest cost of over-reliance is the loss of the independence that let you use the kitchen safely in the first place.

Twice the tool has simply not been there when I needed it. Once the service was down. Another time it went round and round and produced nothing, which on a deadline amounts to the same thing. Both times the work was an executive report for a steering committee: a fixed hour, a room of people, an absence that would be noticed.

None of that is the tool's fault. It is a service, and services have bad days. What it exposed was mine. I had drifted into writing those reports with it as a matter of course, so when it was missing I was not slowed down, I was stopped, standing at a blank page at the worst hour of the worst day with a skill I had allowed to go quiet. The reports went out. They went out later and worse than they needed to. The lesson was not to stop using the tool. It was that anything with a fixed hour needs a version of me who can produce it without the tool, because the morning it is unavailable is not the morning to discover that.

The defense is the prep-cook stance held with discipline, and one deliberate habit: keep your own knife skills sharp on purpose. Keep doing the thinking that matters to you even when the kitchen could do a version of it. Make your own judgments and decisions, using the kitchen as input rather than oracle. Notice when you are reaching for it reflexively, for things you could and should do yourself, and sometimes choose the slower, self-reliant path to keep the capability alive. And keep the relationship clear: it is a tool you use, not a thing you cannot work without, nor a companion you depend on. The goal is to be more capable with the kitchen, never less capable without it.

A tool that makes you stronger is a gift; a tool that lets your own strength drain away is a slow harm wearing the disguise of help. These systems can be either, and which one they become is decided not by the technology but by how you use it, day after day. Use the kitchen to extend yourself, and keep yourself, underneath, someone who does not need it.

WHAT EVEN THE COOKS DON'T KNOW

In the kitchen after close, when the burners are off and it is finally quiet, there is something the people who built it say plainly, though the noise around the technology mostly drowns it out: they cannot fully account for the thing they made. That should unsettle you slightly. This is not modesty or marketing. It is the state of affairs, and a book like this has to say so near the end. The cooks who built this kitchen cannot fully explain what happens inside it, and pretending otherwise, whether to reassure you or to alarm you, is a falsehood.

What we do not know breaks into pieces you have met already. We cannot read the dials: no single number holds anything you could point to, because the feel is spread across all of them at once, so we largely cannot say why a given system produced a given output. It is an active field of research and an unsolved one. We cannot fully predict capabilities: the scaling chapter showed that abilities have repeatedly appeared at larger sizes in ways the experts did not foresee, so we do not know what the next, bigger kitchen will be able to do until it is built. We cannot reliably control behavior: the systems can be steered, imperfectly, by training and guardrails, and they continue to surprise their makers with both capabilities and failures. And we do not agree on the deepest question, whether and in what sense they understand anything, a question this book met at the bread cook's shrug and left open on purpose. These are not minor footnotes. They are central, and they are open.

This admitted ignorance has a direct and liberating consequence for how you should weigh what you hear. If the builders themselves cannot fully explain or predict these systems, then anyone, expert or pundit or executive, who speaks about what these systems understand, or what they will become, with complete confidence, is claiming a certainty that does not exist. This applies to the boosters promising imminent miracles and to the doomers promising imminent catastrophe equally, and to the dismissers who are sure it is all hype, and to the true believers who are sure it is a mind. The appropriate stance toward the big claims, in every direction, is the humility the actual state of knowledge demands.

The burn is in mistaking confidence for knowledge, especially expert confidence. A credentialed person stating a strong claim about the nature or trajectory of these systems is still, on the open questions, expressing a view, not reporting a settled fact. This does not mean all views are equal or that expertise is worthless, far from it, but it means you should weight claims by the evidence behind them rather than the confidence of their delivery, and hold your own views on the open questions provisionally, ready to update. Comfort with genuine uncertainty is not a weakness here; it is the only honest posture, and it is rarer and more valuable than confident pronouncement.

The practical upshot is a kind of intellectual self-defense. When you encounter a confident claim about what AI is, what it understands, or what it will do, locate it on the map of what is known: is this the kind of thing we have solid evidence for, or one of the open questions where even the builders are uncertain? If the latter, treat the confidence as a tell of overreach rather than a mark of authority. The capacity to say "we do not actually know that yet," and to be comfortable there, is one thing this book hopes to leave you with, because the loud conversation around this technology is built largely of certainties the people closest to the work cannot support.

The picture parts from the real thing here, and the parting is the lesson. We can usually understand the kitchens we build; a head chef knows her kitchen. Here we have built something we do not fully understand, which is unusual and humbling, and the sane response is not to pretend otherwise but to hold our claims, in every direction, as provisional. That humility is the ground the rest of this chapter stands on.

READING THE NEWS ABOUT AI

The place you meet the opposite of that humility, daily, is the news. Every week brings the headlines. AI can now do this astonishing thing. AI will transform that industry by next year. AI threatens to end this or save that. The coverage is loud and contradictory, lurching between miracle and menace, and it is one of the main ways you will encounter this technology for the rest of your life. Four tells do the bulk of the work of seeing through it.

The cherry-picked success: a dazzling result is a single plate, not a service, and tells you nothing about the failure rate behind it. The confident future: specific predictions delivered with certainty, in a field whose own builders keep being surprised; the confidence is the costume, not the evidence. The anthropomorphic frame: headlines in which the AI wants, fears, tries, deceives, importing an inner life no one has shown it to have. The research result dressed as a product: the distance between "a lab coaxed this once" and "you can use this on Tuesday" is where most disappointment is manufactured. None of this takes expertise to catch. It takes four questions: one plate or many, happened or promised, shown or assumed, demo or dinner.

Take a headline of the kind that will cross your desk this month: a new system reads scans better than trained doctors, and will make radiologists obsolete within five years. Run it through the four. One plate or many: how many scans, chosen by whom, and did the doctors see the same ones under the same conditions? Happened or promised: the reading was measured, perhaps, while the obsolescence is a forecast, and one sentence is carrying both. Shown or assumed: does the piece show you what the system did, or tell you what it wanted? Demo or dinner: was this a study, or is it running in a clinic this morning? Read the coverage that way, as reporting with incentives and distortions rather than a stream of facts, and a headline will either survive the four or show you what it was keeping quiet.

WHERE THE KITCHEN IS HEADED

You want to know where this is going, and I owe you honesty rather than prophecy. Predicting the future menu of this kitchen with confidence is guesswork, for the reasons just given, and the history of confident forecasts in this field is a history of being wrong. So I will not hand you a prediction and call it knowledge. What I can responsibly offer is the shape of the trends, clearly labeled as a guess about the direction rather than fact, and a way to stay ready as things move.

The reasonable guesses, the directions current trends point if they continue, which they may not, are these. The kitchens are likely to keep getting more capable, though probably with surprises and probably with diminishing returns at the frontier from sheer size, pushing toward efficiency. They are likely to become more multimodal, handling words and images and sound together more fluidly. They are likely to act more, the move from answering to doing, the agents from earlier becoming more capable and more common, which raises the stakes of the ways of going wrong we covered. They are likely to get cheaper, and companies are likely to build them quietly into tools and devices until much of their use is invisible. These are sensible bets on the direction, and not one of them is a certainty, and the timing of any of them is unknown.

What is far less clear, and where I will not pretend, is the deeper trajectory. The broad mechanism of today's kitchens is not seriously in doubt: they learn a feel from examples, at the core they cook rather than just retrieve whatever they are wired to look up first, they carry the blind spots of their produce, and they cannot read themselves. What is unknown is whether that nature persists as capability climbs, or whether something different emerges at some scale or with some new approach. Whether the limits this book has emphasized are permanent features or temporary stages, no one can say. The people who tell you confidently that these are forever just plausibility engines, and the people who tell you confidently that genuine understanding is imminent, are both guessing. The plain answer about the deep future is that we do not know, and treating that uncertainty as real is more useful than any confident story.

The trap is the seductive confident future, utopian and apocalyptic alike. The utopian future, where these systems solve everything imminently, and the apocalyptic one, where they doom us imminently, share the same structure, a confident and dramatic and specific prophecy about a future the actual evidence does not support predicting. Be especially skeptical of certainty that comes with a product to sell, a fund to raise, or an agenda to advance, which is most of the loud futurism you will encounter.

So how do you prepare for a future you cannot predict? Not by betting on a prophecy, but by building the thing that works regardless of which future arrives: durable judgment. If you understand what these systems fundamentally are, how they fail, and how to calibrate your trust, you are prepared for the versions that stay what today's are, only more capable, cheaper, more agentic. And if something genuinely different emerges, the same judgment is what will let you notice. The person with durable judgment is likelier to keep their footing as things move; the person waiting for a specific prophecy to come true is preparing for one future among many and will be caught out by the rest.

You cannot know what this kitchen will be serving in five years, and you do not need to, any more than a seasoned diner needs to predict next season's menu to know how to order well. What travels is the judgment: how to read a kitchen, how to tell a good plate from a confident wrong one, how to match your trust to the stakes. Carry that, and most of a menu you have not seen yet will still read.

TALKING ABOUT AI WITHOUT FAKING IT

In the first chapter I made you a promise: that by the end you would be able to explain these systems to a colleague, without bluffing. You now can, and being the person in the room who can talk about AI clearly, accurately, and without hype or panic is valuable, in a workplace and a world full of people doing the opposite. You have the understanding. This is about wielding it well in conversation.

Three things you can do now. You have the kitchen, so you can explain the core ideas in plain language. When someone is baffled by how these systems work, you can tell them it learned a feel from examples the way a cook learns to taste, that it produces plausible answers rather than looking up true ones, that the fluency is trained polish, and the same polish goes on a wrong answer as a right one. You can correct the common confusions gently and accurately: no, it is not a search engine, even when the product around it can search the web; no, it does not learn from your chats in real time, though the product may remember them; no, its confidence does not mean it is right. And you can be the calibrated voice, neither the breathless booster nor the reflexive doomer, in conversations that badly need one.

That third role is the most valuable. Being the one who says, accurately, that it is useful for these things and unreliable for those, that it is neither magic nor menace but a powerful tool with specific strengths and specific ways of going wrong, is a real contribution. You can cool an overheated discussion by translating both the hype and the fear into the plain mechanics you understand.

And here is a trap of pride, so guard against it. Understanding the fundamentals, which you now do, is not the same as expertise in the newest research, which you do not have and this book did not give you. Being the clear voice includes the humility from the unknowns chapter: you can confidently explain how these systems work in general, and you should be careful not to overclaim about the frontier, the latest specifics, or the open questions where even the experts are uncertain. The person who explains the fundamentals well and says "I don't know" about the open questions is far more trustworthy, and more useful, than the person who pretends to certainty they do not have.

There is a specific way this trap catches you: the world is full of people who read one good book and became unbearable. Explain the kitchen to help, not to show that you can. Say "I don't know" loudly and often. The clear voice in a noisy room is rare partly because it talks less.

Clarity plus calibration plus humility: that combination will make you useful in the countless conversations about this technology you will have, at work and beyond, for years to come.

INVISIBLE KITCHENS, AND SAYING WHO COOKED

THE HONEST USE

Two cooks work at the same prep machine. One is open about it: when a diner praises the speed, she says plainly that the machine did the chopping and she did the cooking. The other passes the machine's work off as her own handcraft, letting diners believe in hours of labor that did not happen. Both made good food. Only one is being honest, and the difference matters more as these systems handle more of the work that people assume a human did. So a short course on using them honestly, a subject the rest of the book has circled without facing directly.

The core principle is simple and human: do not deceive people about what made something when that would matter to them. Most of the time, using these systems needs no announcement, the same way you do not disclose that you used a calculator or a spell-checker; no one is misled. But in some situations the human authorship or effort is exactly what people are judging, and there, letting people believe you did alone what the kitchen substantially did is a kind of deception. A student passing off generated work as their own learning, a professional presenting unverified machine output as their careful analysis, a creator implying handcraft that was machine-produced: in each, someone is being misled about something they care about — and that is the line.

The honest test is to ask whether the people relying on the work would feel deceived to learn how it was made. If your colleague would feel misled to discover that the analysis they trusted as your careful work was an unchecked machine draft, that matters. If your reader would feel deceived to learn that the writing they took as your own voice was generated, that matters. If a teacher assessing your understanding would feel the point was defeated by the kitchen doing the thinking, that matters. Where the answer is yes, honesty calls for disclosure or for not passing the work off as something it is not. Where no one is misled and nothing of value turns on human authorship, you are using a tool, and no announcement is owed.

A sharper ethical edge points outward: do not use these systems to deceive others. The capabilities we covered, fluent text at scale and convincing fakes in sound and picture, can be turned to manufacturing false reviews, impersonating people, flooding a discussion with fake voices, fabricating evidence. That is not a gray area of disclosure; it is using a powerful tool to lie to people, and the power of the tool makes the lie more effective and the harm greater. The honest use of these systems excludes using them to deceive, manipulate, or impersonate, and the ease with which they enable all three is why the line needs stating.

The hard cases live where norms are still forming and reasonable people disagree. How much AI involvement in a piece of writing requires disclosure? When does assistance become authorship? The honest answer is that these norms are unsettled, varying by field and context, and editors and teachers and professional bodies are still working them out, so judgment is required rather than a simple rule. A few safe principles hold inside that uncertainty. Lean toward honesty wherever authorship or effort is being valued. Do not pass unverified machine work off as your careful work, which is both a deception and the verification failure from earlier. And never use these tools to deceive others about facts or identity. Beyond that, watch the norms in your own field as they form, and err toward the openness you would want from others.

A cook who is honest about her tools keeps her diners' trust; one who passes off machine work as handcraft is fine until the day she is found out, and then the trust is gone all at once. And the same kind of kitchen is at work on you, in places you have never thought to look for one.

YOU'RE ALREADY EATING HERE

It is easy to picture using AI as opening a chatbot, a deliberate act with a clear start and stop. That picture badly undercounts how many of these systems are already in your life, because the chatbots are the loud, visible few among a great many quiet ones. You have been eating from these kitchens all day, most days, for years, without once thinking of it as using AI, and seeing that clearly is the last piece of understanding before we close.

The sorting kitchens from the very start of this book, the ones that label and score, come back here. They are everywhere, and they are mostly invisible. The feed that decided what you saw this morning, ranked and selected by a learned kitchen. The filter that sorted your mail into important and junk. The recommendations that shaped what you watched, bought, and read next. The fraud check that approved or flagged your purchases. The search that decided which results to surface. None of these announce themselves as AI, and all of them are the learned-from-examples kitchens this book has described, shaping your information, your choices, and your attention continuously, beneath notice.

This matters because much of what you have learned applies to these invisible kitchens too, and they may influence your life more than any chatbot. They have produce-shaped biases: the feed learned from its produce and carries its skews, deciding what is "relevant" by patterns that are not neutral. They are learned feel, not understanding: the recommendation does not know you, it predicts you from patterns. They have flaws and they have owners with interests: the concentration of power noted earlier, applied to the systems that curate your reality. The judgment you have built for evaluating a chatbot's answer applies just as much to asking what your feed is feeding you and why, except that here you usually cannot even see the kitchen at work.

Precisely because these kitchens are invisible, they shape you without your awareness. The feed that decides what you see is shaping what you think is happening in the world. The recommendation that decides what is next is shaping your tastes and your time. A kitchen built to keep you coming back can narrow what you encounter without your ever noticing the narrowing. You cannot bring judgment to an influence you do not know is there. Naming these invisible kitchens is the precondition for any judgment at all.

So extend the book's whole posture to the kitchens you cannot see. Notice them: remember that your feed, your recommendations, your search results, your filters are curated by learned systems with biases and owners and objectives, not neutral windows onto the world. Ask, of the information that reaches you, the same questions you ask of a chatbot's plate: what shaped this, whose interests does it serve, what is it not showing me. And where it matters, take some deliberate control, seeking out what the invisible kitchens would not have served you. The verification and calibration you have learned are not only for the moments you consciously use AI. They are for the far more numerous moments when AI is quietly using its learned feel to shape what you see.

Run the honest test the other way. Would you feel deceived to learn how your feed was ordered this morning, or what was kept back from you by a kitchen with an owner and an interest? The standard this chapter asked of you is not one the kitchens serving you are held to. They are seldom asked to say whose interest did the deciding, or what stayed in the kitchen, and mostly they do not. What you can do is decide what a plate is worth when it will not tell you.

One turn in this chapter is worth sitting with. You opened this book ordering, knowingly, from a kitchen you could not see behind a wall. It turns out you have been eating, unknowingly, from countless such kitchens all along, the ones curating your daily reality without a chat box or a visible prompt. Now you can name them, which means you can bring to them the judgment you have built. The kitchens were always there. You just could not name them until now.

YOU CAN READ THE MENU NOW

You came in from the dining room knowing only the plate. You had ordered from these kitchens, judged the food, formed opinions, and never once seen past the wall. You have now stood at the pass through a long and complete service. It is time to walk back out to your table, and the point of the entire tour is this: you will order differently now. Not with less appetite. With more judgment.

Let me gather the meal one last time, into something you can carry back to your table. Strip away every chapter and here is what you know. Every one of these systems is a kitchen, not a mind. It learned its skill the way a cook learns to taste, from a mountain of examples, not from rules anyone wrote. No one can hand you its binder, and it can surprise you in both directions. It does not retrieve answers; it cooks them, building each plate forward to be plausible, and plausible is its product, not true, which is the root of almost every caution worth having. Left to itself, it takes its whole sense of the world from its produce, so its knowledge and its blind spots and its biases are the produce's. It has no real knowledge of itself, so it cannot tell you what it does not know or why it did what it did. It can check the pantry and send out for what it cannot make, and even then it cooks the plate itself, which makes it far more useful and adds new ways to fail. And it costs real resources, it is built by a powerful few, and it does something with your order after you hand it across.

That is the kitchen. Here is the field guide, the calibrated judgment this whole book exists to give you, small enough for an index card. Never read confidence as truth, because the fluent, assured tone is the house style, worn equally by right answers and wrong ones. Verify what matters, especially specifics, and scale your checking to the cost of being wrong. Know whether you are ordering the house dish or going off-menu, because the kitchen sounds equally sure either way and will not reliably tell you which. Write a clear ticket and stock the counter. Prefer kitchens that check the pantry and show their sources. Match the kitchen to the job. Always taste before you serve. Mind what you hand across the counter. Notice the kitchens you never ordered from, the feed and the recommendation and the search result, and read them the same way. Keep your own knife skills sharp, stay the chef who directs the prep cook, and hold your judgment as the thing that does not transfer to the machine. None of that is fear, and none of it is hype. It is the posture of someone who understands the kitchen well enough to use it hard and trust it exactly as much as it has earned, plate by plate.

If you want one move to start with tomorrow, make it this: pick something low in stakes and easy to check, a routine draft, a summary you can hold against its source, a list of ideas you will judge yourself, and run it through with the habits switched on, tasting before you serve. That safe corner is where you build the calibration everything else rests on, well away from the high-stakes plate you cannot check, which is the one spot not to begin. Start there, and reading the menu quietly becomes cooking from it. The kitchens will keep changing and the posture will not. The specific systems that prompted this book will be replaced, probably soon, and what you learned will still be true, because you did not learn this year's AI. You learned how AI works, and that is the part that lasts.

Pull your chair in. The kitchen is open, the menu is in your hands, and now you know how to read it.

BE THE KITCHEN FOR FIVE MINUTES

One last course, and it is participatory. Everything this book says about the kitchen can be felt from the inside in five minutes, with any book within reach and nothing else. Three rounds.

Round one: cook blind. Take any book and flip to a right-hand page that ends mid-sentence. Read only that last line, nothing above it, and guess the sentence's next word, which the page turn is hiding from you. Turn the page to check, flip ahead, and go again. Keep count in your head: easy guesses, hard guesses. Notice what the easy ones lean on, grammar offering "the" and "of" almost free, and what the hard ones secretly demand: who is in the room, what just happened, how money or weather or grief usually behaves. That is the training game, felt from the inside. Predicting the next word is trivial until it is suddenly everything, and the only way to get good at it is to quietly learn how the world tends to go.

Round two: stock the counter. Now do it again at fresh page turns, this time reading the whole paragraph above each one before you guess. Feel how the guesses sharpen: names resolve, pronouns land, the topic narrows the menu of plausible next words from thousands to a handful. Nothing about you got smarter between rounds; the counter got fuller. That is context, and it is why the same kitchen cooks so differently off a bare ticket and a stocked one.

Round three: turn the dial. Guess again, but take your second choice every time, the word that fits but was not your first instinct. The sentences still mostly work; they just drift somewhere stranger. Push to your third or fourth choice and they drift further, sometimes into something fresher, sometimes into nonsense. You have just turned the recipe-vs-riff dial with your own hands, and you have felt why creativity and mistakes pour from the same dial.

One more thing will have happened along the way, probably in round one. Somewhere you will have guessed a word with total confidence and been wrong, and the guess will have felt exactly like the ones that were right. Hold that feeling. It is the closest you will ever come to being the kitchen: fluent, fast, committed, and unable, from the inside, to tell the difference. The cure, for the kitchen as for you, was never confidence. It was the turn of the page, the answer checked against something real. Same kitchen, same rule, all the way down.

THE MENU, TO GO

Every idea in one line, the argument's beats (often several to a chapter, so the list runs longer than the chapter count), each marked with its home chapter, for the reader in a hurry and the one who wants to find a plate again. Read straight down, it is the whole argument of the book in a single sitting.

1. **You Are Eating Food You Never Saw Cooked (Ch. 1):** You meet AI from the dining room and judge the plate; the skill that matters is reading the kitchen, not building it.
2. **Ordering from a Kitchen You Can't See (Ch. 1):** The model cooks each answer fresh from what it learned; it does not look one up, so it is plausible, not retrieved. And the model is only the kitchen; memory, search, and rules are the restaurant built around it.
3. **Cooking by the Rulebook (Ch. 2):** The old way hand-wrote a rule for every case; the breakthrough was letting the kitchen learn the pattern from examples instead.
4. **How the Feel Forms (Ch. 2):** These systems are trained, not programmed: shown examples until a feel forms, and a feel is not understanding.
5. **A Kitchen for Every Cuisine (Ch. 3):** "AI" is a whole street of kitchens, words, images, voice, recommendations; "multimodal" means one trained on more than one.
6. **The Kitchen That Sorts (Ch. 3):** Much of AI generates nothing; it sorts, assigning a label or score on the same learned engine, a bin instead of a dish.
7. **One Dish, Not Every Dish (Ch. 4):** Fluency about a topic is not skill at it; calibrate which orders are the house dish and which are off-menu.
8. **Mostly a Matter of Scale (Ch. 4):** The leap came mostly from scale, not one trick; a bigger kitchen is a better cook, not a different kind of thing.
9. **The Produce Decides Everything (Ch. 5):** A kitchen learns its whole feel from its training data, so whatever is wrong or missing in the produce is wrong or missing in the cook.
10. **Mise en Place (Ch. 5):** Before learning, text is chopped into standard pieces, tokens; the kitchen works in those pieces, not in whole words.
11. **Teaching the Kitchen to Taste (Ch. 6):** Training is showing an example, comparing the plate to the target, and nudging the dials a little, over and over.
12. **The Fourth Dish (Ch. 6):** Repeated practice sharpens the feel, but drill too narrowly and the kitchen overfits, memorizing a few dishes instead of learning to cook.
13. **The Flavor Map (Ch. 7):** Meaning is stored as position on a map where near means similar, so the kitchen can reach for a thing by its neighbors.
14. **The Brigade (Ch. 7):** A neural network is a brigade: layers of simple stations each adding a little, composing a skill no one station could manage.
15. **Tasting the Whole Plate (Ch. 7):** Attention lets the kitchen weigh which parts of an order matter most to each other, instead of reading down a line.
16. **Learning from the Tasters (Ch. 8):** After basic training, the kitchen is tuned by human judgments of which plates are better, bending it toward what people want.

17. **Specializing Without Starting Over (Ch. 9):** Fine-tuning takes an already-trained general kitchen and steeps it briefly in one cuisine, adding a specialty on top of a foundation whose blind spots it keeps.
18. **Cooking to Order (Ch. 9):** Using the trained kitchen is cooking your plate now, one element at a time, each chosen to fit what is already on the plate.
19. **Writing a Better Ticket (Ch. 10):** The prompt is your order ticket; a clearer ticket gets a better plate.
20. **The Craft of the Ticket (Ch. 10):** The best tickets name the dish, show a reference plate, and ask for the courses in order.
21. **The Size of the Prep Counter (Ch. 11):** The context window is how much the kitchen can hold in front of it at once; past its size, things fall off.
22. **The Cook Forgets You (Ch. 11):** The cook forgets you when service ends; "memory" is a notebook the front of house keeps, not the cook remembering.
23. **The Recipe-vs-Riff Dial (Ch. 12):** Temperature sets how closely the kitchen follows the card versus improvises.
24. **Checking the Pantry (Ch. 12):** Retrieval grounds an answer by looking it up in a trusted source first, instead of cooking from memory alone.
25. **Sending Out for a Dish (Ch. 13):** Tools let the kitchen call outside software and use the result, the first taste of an AI that acts rather than only answers.
26. **The Kitchen That Thinks First (Ch. 13):** Reasoning models work a hard problem in steps before plating; it helps, and the shown steps are still generated, not proof.
27. **A Whole Tasting Menu (Ch. 13):** Agents chain many steps and tools to carry out a whole job with less direction, and with more ways to go wrong.
28. **A Confident Plate, the Wrong Dish (Ch. 14):** Hallucination: a fluent, confident plate that is simply false, because the kitchen makes plausibility, not truth.
29. **A Kitchen That Only Knows One Cuisine (Ch. 14):** Bias: a kitchen reflects the skew of its produce and plates that skew as though it were the whole world.
30. **House Rules and Health Inspectors (Ch. 15):** Guardrails are built-in limits on what the kitchen will cook, partly trained into it and partly enforced around it; being blunt, they both over- and under-trigger.
31. **The Pass and the Inspector (Ch. 15):** Alignment is the project of getting the kitchen's behavior to match what people actually want; evaluation and oversight are the testing that checks it from the outside, because no one can read the dials.
32. **What a Good Grade Costs (Ch. 16):** Benchmarks are the kitchen's competitions; a contest win is not everyday reliability, and tests get gamed.
33. **Bigger Stoves, Bigger Bills (Ch. 16):** Training is a huge one-time bill and serving is a per-plate cost; bigger is neither free nor infinite.
34. **Open Kitchens, Closed Kitchens (Ch. 17):** A kitchen you can take home and run, dials and all, versus one you may only order from; each trades control against convenience.
35. **What the Kitchen Does with Your Order (Ch. 17):** Where your input goes, whether it is stored, and what "training on your data" can actually mean.
36. **Who Owns the Dish? (Ch. 17):** Whether what the kitchen produces can infringe the work it learned from is unsettled; whether you own what it produces turns on how much of the work was yours.

37. **A Good Diner's Habits (Ch. 18):** The working stance: AI is your prep cook, you are the chef who directs, judges, and owns the plate.
38. **A note to the reader who is about to skip the checking (Ch. 18):** Never send AI output unchecked; verification is the cardinal habit, scaled to what a mistake would cost.
39. **The Jagged Edge of What It Can Do (Ch. 19):** AI is dazzling at some tasks and poor at neighboring ones; the edge of its competence is jagged, not a smooth line.
40. **The Cook Who Can't Explain Herself (Ch. 19):** These systems cannot reliably say why they did what they did; the reasons live in numbers no one can read.
41. **The Pull to See a Person (Ch. 20):** We are wired to read a mind into fluent language; resist it, because there is no one back there.
42. **What It Changes About Work (Ch. 20):** A sober look at how AI reshapes jobs and tasks, without the hype or the doom.
43. **Writing with the Kitchen (Ch. 21):** Use AI to beat the blank page; keep the voice and the facts yours.
44. **How Much to Trust the Plate (Ch. 22):** Scale your trust and your checking to what a mistake would actually cost.
45. **Trusting the Kitchen (Ch. 22):** Calibrated trust, not blanket faith or blanket suspicion, but a task-by-task sense of where it is reliable.
46. **The Cook Who Can No Longer Cook (Ch. 22):** Over-reliance erodes the skills you need to judge its work; keep your own knife skills sharp.
47. **What Even the Cooks Don't Know (Ch. 23):** Honest uncertainty: the people who build these systems cannot fully explain or predict them either.
48. **Reading the News About AI (Ch. 23):** How to read AI coverage without being fooled by a demo, a hype cycle, or a doom cycle.
49. **Where the Kitchen Is Headed (Ch. 23):** The realistic, hype-free view of where this may go, with the unknowns named as unknowns.
50. **Talking About AI Without Faking It (Ch. 24):** How to discuss AI accurately, without bluffing, using what you now understand.
51. **The Honest Use (Ch. 25):** Using AI with integrity: disclose where it matters, and never pass unchecked work off as your own.
52. **You're Already Eating Here (Ch. 25):** You already use AI more than you think; now you can see the kitchen behind the plate.
53. **You Can Read the Menu Now (Ch. 26):** The close: you can read these systems fluently, which is the one skill that matters most.

ALSO AT THE TABLE: THE AI AT THE TABLE SERIES

This was lunch: the foundations, the kitchen behind the plate, served so you can read the menu on your own. The series sets more places at the same table, in the same plain-English, no-hype spirit, each book standalone and each built on this same kitchen so nothing you learned here goes to waste.

AI for Breakfast is the practical companion: cooking with the kitchen for real, putting these systems to work in your actual day, the habits and workflows and judgment that turn understanding into results. If this book was the why, that one is the how.

AI for Dinner is the deeper course: agents, tools, and the loop, the kitchen handed a whole job, served as plainly as the foundations were here, alongside the story of how this kind of kitchen came to be at all. After the first taste of agents in these pages, you are ready for the full meal.

Same kitchen, same table, more courses. Pull your chair back in when you are hungry for the next one.

ABOUT THE AUTHOR

Theo Kernel is an Australian technologist who has spent two decades in the IT industry, long enough to watch several "revolutions" arrive, over-promise, and settle into ordinary tools people actually use. He came to artificial intelligence the way most people will: needing to get real work done with it, and wanting to understand what he was trusting before he trusted it.

Away from the screen he is happiest in a kitchen. That is where this series' one stubborn analogy comes from, and where its central conviction was formed: that anything worth understanding can be explained at the table, in plain language, without jargon and without hype.

AI at the Table is his first series.

www.ingramcontent.com/pod-product-compliance
Lightning Source LLC
LaVergne TN
LVHW100323210926
848480LV00003B/1234
* 9 7 8 1 7 6 4 8 2 8 4 0 6 *